Lecture Notes in Mathematics

Edited by A. Dold and B. Eckmann

872

Set Theory and Model Theory

Proceedings of an Informal Symposium
Held at Bonn, June 1–3, 1979

Edited by R. B. Jensen and A. Prestel

Springer-Verlag
Berlin Heidelberg New York 1981

Editors

Ronald Björn Jensen
All Souls College
Oxford OX1 4AL, England

Alexander Prestel
Fakultät für Mathematik, Universität Konstanz
Postfach 5560, 7750 Konstanz,
Federal Republic of Germany

AMS Subject Classifications (1980): 03 Cxx, 03 Exx

ISBN 3-540-10849-1 Springer-Verlag Berlin Heidelberg New York
ISBN 0-387-10849-1 Springer-Verlag New York Heidelberg Berlin

Printing and binding: Beltz Offsetdruck, Hemsbach/Bergstr.
2141/3140-543210

FOREWORD

On the occasion of Gisbert Hasenjaeger's 60-th birthday an informal symposium on set theory and model theory was held at Bonn from 1.-3. of June 1979. The papers published in these proceedings are all dedicated to Professor Hasenjaeger. Since the meeting took place some of the papers have been revised and extended. All of the contributors to this volume are former students or co-workers of Professor Hasenjaeger. Each was at one time or another his assistant.

Each of us, in his own approach to mathematics, bears deep traces of Hasenjaeger's influence. Hasenjaeger himself is an eager proponent of the platonic view advocated by his teacher, Heinrich Scholz. Though not all of us would call ourselves platonists today, we retain a profound sense that mathematics must have a 'real content' - and that a problem should be judged by its content, rather than its difficulty or esthetic value. All of us learned from Hasenjaeger to regard the concept of model as central in logic - a conviction to which this book bears witness.

The preponderance of set theoretical contributions to this volume is related directly to Hasenjaeger's insistence on the importance of ontological questions. We fear, however, that Hasenjaeger, were he not too polite to do so, would accuse the set theorists among us of 'taking the easy road'. He recognizes the primacy of the natural numbers and feels that a major mathematical direction should be able to shed light on them. In particular, he has faith that the tools of logic will some day prove the elementary undecidability of some of the great unsolved questions of number theory (and in the process perhaps even point the way to their solution in some stronger theory). For this reason he has made a lifelong study of non-standard models and encouraged those around him to work on them. Some of us tried, but all eventually turned to fields with more immediate rewards. Recently, however, the beautiful theorem of Paris and Harrington provided a first vindication of his faith. We hope that more will follow.

We are grateful to Professor Hasenjaeger for the many human qualities - his humor, his tolerance, his patient attentiveness - which made our association with him so pleasant.

R. B. Jensen

A. Prestel

FOREWORD

[illegible] of Professor Hasenjaeger [illegible] on [illegible] theory and model theory [illegible] The papers published [illegible] also [illegible] Professor Hasenjaeger [illegible] have been revised [illegible] volume are [illegible] Hasenjaeger's [illegible]

With [illegible] his own approach to mathematics [illegible] Professor Hasenjaeger's [illegible] that a problem should be judged [illegible] — a [illegible] which this book [illegible].

The [illegible] of set theoretical [illegible] this volume is related [illegible] Hasenjaeger [illegible] we [illegible] he has made a [illegible] study of [illegible] and [illegible] him to work on them. Some of us [illegible] with more [illegible]. Recently, however, [illegible] theorem [illegible] provided a first [illegible]. We hope that more will follow.

We are grateful to Professor Hasenjaeger for [illegible] many human qualities [illegible] his tolerance, his [illegible] which made our association with him so pleasant.

R. [illegible]

A. [illegible]

TABLE OF CONTENTS

Morass-Like Constructions of $\aleph_2$-Trees in L

by

Keith J. Devlin

(Lancaster, U.K.)

Abstract

Using a simplified version of the fine structure theory required for the construction of a morass, we show how to construct Souslin and Kurepa $\aleph_2$-trees as limits of directed systems of countable trees. These techniques were originally developed by us in order to obtain a Souslin $\aleph_2$-tree T such that the reduced power tree T^{ω}/D, where D is a non-principal, uniform filter on ω, is a Kurepa $\aleph_2$-tree. However, subsequent to our work, R. Laver obtained a much simpler proof, so we do not include this result here. Rather the present account is largely expository in nature.

§ 1. Preliminaries

We work in ZFC set theory and adopt the usual notation and conventions. In particular, ordinals are von Neumann ordinals and cardinals are initials ordinals. We use α, β, γ, etc. to denote ordinals; $|X|$ denotes the cardinality of the set X.

Most of our results require the axiom of constructibility, V = L, and we refer to our monograph [De 1] for details. Section 2 provides an outline of the morass theory we need. It turns out that we do not require the full power of the morass (in particular, morass axiom (M7) from [De 1] is not required), but we do need to know something of the construction of the morass (in L).

Our terminology concerning trees is fairly standard, but in view of the importance of the various definitions to this paper we present here a quick resumé of what we need.

A tree is a poset $\underset{\sim}{T} = \langle T, \leq_T \rangle$ such that the set $\hat{x} = \{y \in T | y <_T x\}$ is well-ordered by $<_T$ for each x in T. The order-type of $\hat{x}$ (under $<_T$) is called the height of x in $\underset{\sim}{T}$, denoted by ht(x). The α'th level of $\underset{\sim}{T}$ is the set $T_\alpha = \{x \in T | ht(x) = \alpha\}$. The height of $\underset{\sim}{T}$ is the least λ such that $T_\lambda = \phi$, and is denoted by $ht(\underset{\sim}{T})$. The width of $\underset{\sim}{T}$ is the least cardinal κ such that $(\forall\alpha)\ (|T_\alpha| < \kappa)$. For $\alpha < ht(\underset{\sim}{T})$, we set $T\restriction\alpha = \bigcup_{\beta<\alpha} T_\beta$. $\underset{\sim}{T}\restriction\alpha$ denotes the restriction of $\underset{\sim}{T}$ to the set $T\restriction\alpha$, though in general we do not distinguish between a tree and its domain.

Let $\underset{\sim}{T}$ be a tree. A branch of $\underset{\sim}{T}$ is a maximal, linearly ordered subset of $\underset{\sim}{T}$. A branch whose order-type is α (under $<_T$) is called an α-branch. An antichain of $\underset{\sim}{T}$ is a pairwise incomparable subset of $\underset{\sim}{T}$.

A tree $\underset{\sim}{T}$ is normal iff:

(i) if $\alpha < \beta < ht(\underset{\sim}{T})$, every element of T_α has infinitely many extensions on T_β;

(ii) if $\alpha < ht(\underset{\sim}{T})$ is a limit ordinal, any α-branch of $T\restriction\alpha$ has at most one extension on T_α.

An $\aleph_2$-tree is a normal tree T of height ω_2 and width $\aleph_2$.

An $\aleph_2$-tree is Aronszajn iff it has no ω_2-branch. Assuming CH, it is easy to construct an Aronszajn $\aleph_2$-tree. (See [Je].). Silver has shown that the assumption of CH is necessary here. (See [Mi].)

An $\aleph_2$-tree is Souslin iff it has no antichain of cardinality $\aleph_2$. Assuming V = L, it is easy to construct a Souslin $\aleph_2$-tree. (See [De 1].) Laver ([La]) has shown that CH is not sufficient here. It is easily seen that every Souslin $\aleph_2$-tree is Aronszajn.

An $\aleph_2$-tree is Kurepa iff it has at least $\aleph_2$ many ω_2-branches. Assuming V = L, it is easy to construct Kurepa $\aleph_2$-trees. (See [De 1].) Silver ([Si]) has shown that CH is not sufficient for this.

The following definitions are non-standard. If A is a subset of a tree T, we set

$$\hat{A} = \{x \in T \mid (\exists a \in A)(x \leq_T a)\},$$

the initial part of T determined by A. A subset U of a tree T is said to be thin if

$$(\forall x \in U)(\exists y \in T)(x <_T y \ \& \ y \notin U).$$

§ 2. Some Morass Theory

The concept of a morass was introduced by Ronald Jensen. Amongst other results, he used morasses to prove the Gap-2 Two Cardinals Conjecture of model theory under the assumption V = L. In Chapter 13 of our monograph [De 1] we show how to construct a morass in L. Unfortunately, for our present purposes, it is not enough simply to have the definition of a morass, and the construction given in [De 1] is not adequate for us. Rather we require a proper fragment of an alternative (and vastly superior) construction of a morass. This alternative construction will be given in full in the forthcoming book [De 2], but in view of the fact that this volume is some way off completion, we give here the theory we require. This serves the additional purpose of committing to print at least part of the "new" morass construction.

We assume the reader is familiar with the fine structure theory as described in Chapters 6 and 7 of [De 1]. We use the same notation as there. Except for this (standard) material, the following account is self-contained. The reader can ignore the construction of the morass in [De 1] (Chapter 13). We assume V = L from now on.

Let us call an ordinal $\nu < \omega_2$ <u>suitable</u> iff:

(i) either ν is admissible or else $\{\tau \in \nu \mid \tau$ is admissible$\}$ is unbounded in ν; and

(ii) $J_\nu \models$ "there is exactly one uncountable cardinal."

Let S be the set of all suitable ordinals. For $\nu \in S$, set $\alpha_\nu = \omega_1^{J_\nu}$.

Let

$$A = \{\alpha_\nu \mid \nu \in S\},$$

and for $\alpha \in A$, set

$$S_\alpha = \{\nu \in S \mid \alpha_\nu = \alpha\}.$$

The following lemma is easily proved:

<u>Lemma 2.1</u>

(i) $\omega_1 \in A$;

(ii) $A \cap \omega_1$ is unbounded in ω_1;

(iii) S_{ω_1} is closed and unbounded in ω_2;

(iv) S_α is closed in $\sup(S_\alpha)$;

(v) if $\bar{\alpha}, \alpha \in A$ and $\bar{\alpha} < \alpha$, then $\max(S_{\bar{\alpha}}) < \alpha$;

(vi) for any $\nu \in S$, $S_{\alpha_\nu} \cap \nu$ is uniformly Σ_1-definable from α_ν in J_ν. □

Let ν be any limit ordinal. For $\beta \geq \nu$, we say ν is <u>singular over</u> J_β iff there is a function f, definable as a class over J_β, which maps a bounded subset of ν cofinally into ν. We say ν is Σ_n-<u>singular over</u> J_β if there is such an f which is definable over J_β by means of a Σ_n-formula of set theory (with parameters from J_β). Otherwise ν is <u>regular over</u> J_β or Σ_n-<u>regular over</u> J_β, respectively.

Let $\nu \in S$. Then ν is singular. Hence there is (by V = L) a $\beta \geq \nu$ such that ν is singular over J_β. Let $\beta(\nu)$ be the least such β. Let $n(\nu)$ be the least integer $n \geq 1$ such that ν is Σ_n-singular over $J_{\beta(\nu)}$. Define

$$\rho(\nu) = \rho_{\beta(\nu)}^{n(\nu)-1}, \qquad A(\nu) = A_{\beta(\nu)}^{n(\nu)-1}.$$

Since ν is $\Sigma_{n(\nu)-1}$-regular over $J_{\beta(\nu)}$, we have $\rho(\nu) \geq \nu$. If, in fact, $\rho(\nu) < \nu$, then since $\rho(\nu) \leq \beta(\nu)$, we have $J_{\rho(\nu)} \models$ "ν is regular". Using the fact that $\alpha_\nu = \omega_1^{J_\nu}$ in case $\rho(\nu) > \nu$, it follows that in all cases, $J_{\rho(\nu)} \models$ "α_ν is regular". Notice also that if $\tau \in S_{\alpha_\nu} \cap \nu$, then since α_ν is the largest cardinal in J_ν, τ is not a cardinal in J_ν, so $\beta(\tau) < \nu$, which implies that $\rho(\tau) < \nu$.

We now define, for each $\nu \in S$, a certain parameter $p(\nu) \in J_{\rho(\nu)}$. The definition of $p(\nu)$ depends upon the nature of ν, and there are two cases to consider.

Let $\quad P = \{\nu \in S \mid n(\nu) = 1 \ \& \ \beta(\nu) \text{ is a successor ordinal}\}$,

$\quad R = S - P$.

In case $\nu \in P$, let $\gamma(\nu)$ be that γ such that $\gamma + 1 = \beta(\nu)$.

<u>Lemma 2.2</u>

For any $\nu \in S$, $\rho^{n(\nu)}_{\beta(\nu)} \leq \alpha_\nu$.

Proof: Let $n = n(\nu)$, $\beta = \beta(\nu)$, $\alpha = \alpha_\nu$. Let $f \subseteq \nu \times \alpha$ be a $\Sigma_n(J_\beta)$ map such that $f[\alpha]$ is cofinal in ν. Either $\beta = \nu$ or else ν is regular in J_β, so we must have $f \notin J$. But if $\rho^n_\beta > \nu$, then by amenability, $f = f \cap (\nu \times \nu) \in J_{\rho^n_\beta} \subseteq J_\beta$. Hence $\rho^n_\beta \leq \nu$, and it follows that there is a $\Sigma_n(J_\beta)$ map g such that $g[\nu] = J_\beta$. Now let $\langle A_\xi \mid \xi < \alpha \rangle$ be a partition of α into α many sets of cardinality α in J_ν, and let f_ξ be the $<_J$-least map of A_ξ onto $f(\xi)$ for each $\xi \in \mathrm{dom}(f)$. Set $k = \bigcup\{f_\xi \mid \xi \in \mathrm{dom}(f)\}$. Then k is $\Sigma_n(J_\beta)$ and $k[\alpha] = \nu$. Hence $g \circ k$ is $\Sigma_n(J_\beta)$ and $g \circ k[\alpha] = J_\beta$. This proves the lemma. $\square$

<u>Case 1</u>. $\nu \in P$.

Set $\alpha = \alpha_\nu$, $\beta = \beta(\nu)$, $n = n(\nu)$, $\rho = \rho(\nu)$, $A = A(\nu)$, $\gamma = \gamma(\nu)$.

<u>Lemma 2.3</u>

There is a $q \in J_\gamma$ such that every element of J_γ is J_γ-definable from parameters in $J_\alpha \cup \{q\}$.

Proof: By lemma 2.2 there is a $p \in J$ such that every element of J_β is Σ_1-definable in J_β from parameters in $J_\alpha \cup \{p\}$. Since $J_\beta = \mathrm{rud}(J_\gamma)$ there is a rudimentary function f and an element q of J_γ such that $p = f(J_\gamma, q)$. We show that q is as above.

Let $x \in J_\gamma$. Then for some Σ_0 formula ϕ and some $\vec{z} \in J_\alpha$, x is the unique x in J_β such that $(\exists y \in J_\beta)\ [J_\beta \models \phi(y, \vec{z}, p, x)]$. Since $J_\beta = \bigcup_{m<\omega} \underset{\sim}{S}^m(J_\gamma)$, we can find an $m < \omega$ such that

$$(\exists y \in \underset{\sim}{S}^m(J_\gamma))\ [\underset{\sim}{S}^m(J_\gamma) \models \phi(y, \vec{z}, p, x)].$$

Since $p = f(J_\gamma, q)$, this can be written as a first-order statement in J_γ, with parameters $x, \vec{z}, q$. Hence x is J_γ-definable from $\vec{z}, q$. $\square$

Let $q = q(\nu)$ be the $<_J$-least $q \in J_\gamma$ such that every element of J_γ is J_γ-definable from parameters in $\alpha \cup \{q\}$.

Set $p(\nu) = \langle q,\gamma,\nu,\alpha\rangle$. That completes the definitions in Case 1.

Case 2. $\nu \in R$.

By lemma 2.2, let $q(\nu)$ be the $<_J$-least $q \in J_{\rho(\nu)}$ such that every element of $J_{\rho(\nu)}$ is Σ_1-definable from parameters in $\alpha_\nu \cup \{q\}$ in $\langle J_{\rho(\nu)},A(\nu)\rangle$. Set

$$p(\nu) = \begin{cases} \langle q(\nu),\nu,\alpha_\nu\rangle, & \text{if } \nu < \rho(\nu), \\ \langle q(\nu),\alpha_\nu\rangle, & \text{if } \nu = \rho(\nu). \end{cases}$$

This completes the definition in Case 2.

The following two lemmas are clear:

Lemma 2.4

$\langle\langle\beta(\eta),\rho(\eta),A(\eta),p(\eta)\rangle \mid \eta \in S_{\alpha_\nu} \cap \nu\rangle$ is uniformly $\Sigma_1^{J_\nu}(\{\alpha_\nu\})$ for $\nu \in S$. □

Lemma 2.5

Let $\nu,\tau \in S$, and suppose

$$\sigma\colon \langle J_{\rho(\nu)},A(\nu)\rangle \prec_1 \langle J_{\rho(\tau)},A(\tau)\rangle$$

is such that $\sigma(p(\nu)) = p(\tau)$.

Then σ is uniquely determined by $\sigma \restriction \alpha_\nu$. Moreover:

(i) $\nu \in P \leftrightarrow \tau \in P$;

(ii) $\sigma(\alpha_\nu) = \alpha_\tau$;

(iii) $\nu < \rho(\nu) \leftrightarrow \tau < \rho(\tau)$;

(iv) $\nu < \rho(\nu) \rightarrow \sigma(\nu) = \tau$;

(v) $\nu \in P \rightarrow \sigma(\gamma(\nu)) = \gamma(\tau)$;

(vi) $\sigma(q(\nu)) = q(\tau)$. □

Lemma 2.6

Let $\nu \in S$, $\bar{\rho} \leq \rho(\nu)$, $\bar{A} \subseteq J_{\bar{\rho}}$, and suppose

$$\sigma: \langle J_{\bar{\rho}}, \bar{A} \rangle \prec_1 \langle J_{\rho(\nu)}, A(\nu) \rangle,$$

where $p(\nu) \in \text{ran}(\sigma)$. Then there is a (necessarily unique) $\bar{\nu} \in S$ such that $\bar{\rho} = \rho(\bar{\nu})$; $\bar{A} = A(\bar{\nu})$. Moreover, $\sigma(p(\bar{\nu})) = p(\nu)$.

Proof: Notice first that $\langle J_{\bar{\rho}}, \bar{A} \rangle$ is amenable. For if $\nu \in P$, then $A(\nu) = \phi$, so $\bar{A} = \phi$, and if $\nu \in R$, then $\lim(\rho(\nu))$, so $\lim(\bar{\rho})$, and for each $\eta < \bar{\rho}$,

$$\langle J_{\rho(\nu)}, A(\nu) \rangle \models \exists x\ (x = A(\nu) \cap J_{\sigma(\eta)}),$$

so

$$\langle J_{\bar{\rho}}, \bar{A} \rangle \models \exists x\ (x = \bar{A} \cap J_{\eta}).$$

Set $\alpha = \alpha_\nu$, $\beta = \beta(\nu)$, $n = n(\nu)$, $\rho = \rho(\nu)$, $A = A(\nu)$, $p = p(\nu)$, $q = q(\nu)$.

Case 1. $\nu \in P$

Set $\gamma = \gamma(\nu)$. Thus $\beta = \rho = \gamma + 1$, $\bar{A} = A = \phi$, and ν is regular over J_γ. Since $p \in \text{ran}(\sigma)$, we have $q, \gamma, \nu, \alpha \in \text{ran}(\sigma)$. Let $\bar{q} = \sigma^{-1}(q)$, $\bar{\gamma} = \sigma^{-1}(\gamma)$, $\bar{\nu} = \sigma^{-1}(\nu)$, $\bar{\alpha} = \sigma^{-1}(\alpha)$. Let $\bar{\sigma} = \sigma \upharpoonright J_{\bar{\gamma}}$. Thus $\bar{\sigma}: J_{\bar{\gamma}} \prec J_\gamma$ and $q \in \text{ran}(\bar{\sigma})$.

Claim A: $\bar{\nu} \in S$ and $\alpha_{\bar{\nu}} = \bar{\alpha}$.

Proof: This follows immediately from $\sigma: J_{\bar{\rho}} \prec_1 J_\rho$ and $\sigma(\bar{\nu}) = \nu$, $\sigma(\bar{\alpha}) = \alpha$.□

Claim B: $\bar{\nu}$ is regular over $J_{\bar{\gamma}}$.

Proof: Since $\bar{\sigma}: J_{\bar{\gamma}} \prec J_\gamma$ and $\bar{\nu} = \bar{\gamma} \to \nu = \gamma$ and $\bar{\nu} < \bar{\gamma} \to \bar{\sigma}(\bar{\nu}) = \nu$.□

Claim C: $\bar{q}$ is the $<_J$-least element of $J_{\bar{\gamma}}$ such that every element of $J_{\bar{\gamma}}$ is $J_{\bar{\gamma}}$-definable from parameters in $\bar{\alpha} \cup \{\bar{q}\}$.

Proof: Let $x \in J_{\bar{\gamma}}$. Then $\bar{\sigma}(x) \in J_\gamma$, so for some $\vec{\delta} \in \alpha$, $\bar{\sigma}(x)$ is J_γ-definable from $q, \vec{\delta}$. Set $s = \langle \vec{\delta} \rangle$. Let ϕ be a formula of set theory such that

(i) $J_\gamma \models \forall z\ \exists y\ \forall y'\ [y' = y \leftrightarrow \phi(y',z,q)]$;

(ii) $J_\gamma \models \forall z\ \forall y\ [\phi(y,z,q) \rightarrow (\exists \vec{\xi})\ (z = \langle\vec{\xi}\rangle)]$;

(iii) $(\forall y \in J_\gamma)\ [y = \bar{\sigma}(x) \leftrightarrow J_\gamma \models \phi(y,s,q)]$.

Let t be the $<_J$-least element of J_γ such that $J_\gamma \models \phi(\bar{\sigma}(x),t,q)$. Then t is J_γ-definable from $\bar{\sigma}(x)$,q. But $\bar{\sigma}(x), q \in \mathrm{ran}(\bar{\sigma}) \prec J_\gamma$. Hence $t \in \mathrm{ran}(\bar{\sigma})$. By choice of t, $t \leq_J s$, so $t \in J_\alpha$. Thus $t = \langle\vec{\xi}\rangle$ for some $\vec{\xi} \in \alpha$.

By (i) above,

$$(\forall y \in J_\gamma)\ [y = \bar{\sigma}(x) \leftrightarrow J_\gamma \models \phi(y,t,q)].$$

Applying $\bar{\sigma}^{-1}$ and setting $\bar{t} = \bar{\sigma}^{-1}(t) = \langle\vec{\zeta}\rangle$, we get

$$(\forall y \in J_{\bar{\gamma}})\ [y = x \leftrightarrow J_{\bar{\gamma}} \models \phi(y,\bar{t},\bar{q})].$$

Since $\vec{\zeta} \in \bar{\alpha}$, this means that x is $J_{\bar{\gamma}}$-definable from parameters in $\bar{\alpha} \cup \{\bar{q}\}$.

Hence every element of $J_{\bar{\gamma}}$ is $J_{\bar{\gamma}}$-definable from parameters in $\bar{\alpha} \cup \{\bar{q}\}$. We show that $\bar{q}$ is $<_J$-minimal with this property. Suppose not, and let $\bar{q}' <_J \bar{p}$ have this property also. Then, in particular, there are $\vec{\zeta} \in \bar{\alpha}$ such that $\bar{q}$ is $J_{\bar{\gamma}}$-definable from $\vec{\zeta},\bar{q}'$. Applying $\bar{\sigma}: J_{\bar{\gamma}} \prec J_\gamma$ and setting $q' = \bar{\sigma}(\bar{q}')$, $\vec{\xi} = \bar{\sigma}(\vec{\zeta})$, we have $q' <_J q$ and q is J_γ-definable from $\vec{\xi}$,q'. Hence every element of J_γ is J_γ-definable from parameters in $\bar{\alpha} \cup \{q'\}$, since in any J_γ-definition of an element from parameters in $\bar{\alpha} \cup \{q\}$ we can replace q by its definition from $\vec{\xi}$,q'. This contradicts the definition of q.□

Claim D: $\bar{\nu}$ is not Σ_1-regular over $J_{\bar{\gamma}+1}$.

Proof: For each $m \in \omega$, set

$$X_m = \{x \in J_{\bar{\gamma}} \mid x \text{ is } \Sigma_{m+1}\text{-definable from parameters in } \bar{\alpha} \cup \{\bar{q}\} \text{ in } J_{\bar{\gamma}}\}.$$

Then $X_m \prec_m J_{\bar{\gamma}}$ and there is a $J_{\bar{\gamma}}$-definable map of $\bar{\alpha}$ onto X_m. Since $\bar{\alpha}$ is the largest cardinal in $J_{\bar{\nu}}$ and $\bar{\alpha} \subseteq X_m \prec_1 J_{\bar{\gamma}}$, $X_m \cap \bar{\nu}$ is transitive, so set $\bar{\nu}_m = X_m \cap \bar{\nu}$. Since $\bar{\nu}$ is regular over $J_{\bar{\gamma}}$ and there is a $J_{\bar{\gamma}}$-definable map of $\bar{\alpha}$ onto $\bar{\nu}_m$, we have $\bar{\nu}_m < \nu$. But by claim C, $\bigcup_{m<\omega} X_m = J_{\bar{\gamma}}$, so $\sup_{m<\nu}\bar{\nu}_m = \nu$. Clearly, $\langle\bar{\nu}_m \mid m<\omega\rangle$ is $\Sigma_1(J_{\bar{\gamma}+1})$, so we are done. □

Claim E: $\beta(\bar{\nu}) = \bar{\gamma} + 1$, $n(\bar{\nu}) = 1$, $\bar{\nu} \in P$, $\rho(\bar{\nu}) = \bar{\gamma} + 1$, $\gamma(\bar{\nu}) = \bar{\gamma}$.

Proof: By claims B and D. □

Claim F: $q(\bar{\nu}) = \bar{q}$ and $\sigma(p(\bar{\nu})) = p(\nu)$.

Proof: By claims C and E, $q(\bar{\nu}) = \bar{q}$. So by claims A and E, $p(\bar{\nu}) =$ $= \langle\bar{q},\bar{\gamma},\bar{\nu},\bar{\alpha}\rangle$, which implies $\sigma(p(\bar{\nu})) = p(\nu)$. □

That completes the proof of the lemma in Case 1.

Case 2. $\nu \in R$.

Set $\bar{q} = \sigma^{-1}(q)$, $\bar{\alpha} = \sigma^{-1}(\alpha)$. Set $\bar{\nu} = \sigma^{-1}(\nu)$ if $\nu < \rho$ and set $\bar{\nu} = \bar{\rho}$ if $\nu = \rho$. By the fine structure theory ([De 1], page 100) there is a unique $\bar{\beta} \geq \bar{\rho}$ such that $\bar{\rho} = \rho_{\bar{\beta}}^{n-1}$, $\bar{A} = A_{\bar{\beta}}^{n-1}$, and an embedding $\tilde{\sigma}: J_{\bar{\beta}} \prec_n J_\beta$ such that $\sigma \subseteq \tilde{\sigma}$.

Claim G: $\bar{\nu} \in S$ and $\alpha_{\bar{\nu}} = \bar{\alpha}$.

Proof: If $\nu = \rho$, then $\bar{\nu} = \bar{\rho}$ and $\sigma: J_{\bar{\nu}} \prec_1 J_\nu$. And if $\nu < p$, then $\sigma(\bar{\nu}) = \nu$, so $(\sigma \restriction J_{\bar{\nu}}): J_{\bar{\nu}} \prec J_\nu$. Moreover, $\sigma(\bar{\alpha}) = \alpha$. The claim follows immediately. □

Claim H: $\bar{\nu}$ is Σ_{n-1}-regular over $J_{\bar{\beta}}$.

Proof: Suppose not. Then, since $\bar{\alpha}$ is the largest cardinal in $J_{\bar{\nu}}$, we can find a $\Sigma_{n-1}(J_{\bar{\beta}})$ map f such that $f[\bar{\alpha}]$ is cofinal in $\bar{\nu}$. There are now two cases to consider.

Suppose first that $\nu < \rho$. Thus $\bar{\nu} < \bar{\rho}$ and $\sigma(\bar{\nu}) = \nu$. If $f \in J_{\bar{\beta}}$, then by applying $\tilde{\sigma}: J_{\bar{\beta}} \prec_n J_\beta$ we see that $\tilde{\sigma}(f)$ maps a subset of α cofinally into ν, which is impossible since ν is regular in J_ρ. Thus $f \notin J_{\bar{\beta}}$. By using Gödel's pairing function we can code f as a $\Sigma_{n-1}(J_{\bar{\beta}})$ subset of

$\bar{\nu}$, now, to conclude that $\mathcal{P}(\bar{\nu}) \cap \Sigma_{n-1}(J_{\bar{\beta}}) \not\subseteq J_{\bar{\beta}}$. Thus $\rho_{\bar{\beta}}^{n-1} \leq \bar{\nu}$, contrary to $\rho_{\bar{\beta}}^{n-1} = \bar{\rho} > \bar{\nu}$.

Suppose now that $\nu = \rho$. Thus $\bar{\nu} = \bar{\rho}$. In $J_{\bar{\nu}}$, let $\langle A_\xi \mid \xi < \bar{\alpha}\rangle$ be a partition of $\bar{\alpha}$ into $\bar{\alpha}$ many sets of cardinality $\bar{\alpha}$. For each $\xi \in \mathrm{dom}(f)$, let $k_\xi \in J_{\bar{\nu}}$ be the $<_J$-least map of A_ξ onto $f(\xi)$. Set $k = \bigcup\{k_\xi \mid \xi \in \mathrm{dom}(f)\}$. Clearly, k is a $\Sigma_{n-1}(J_{\bar{\beta}})$ function such that $k[\bar{\alpha}] = \bar{\nu}$. But $\bar{\nu} = \bar{\rho} = \rho_{\bar{\beta}}^{n-1}$ and $\bar{\alpha} < \bar{\nu}$, so this is impossible. □

<u>Claim I:</u> $\bar{q}$ is the $<_J$-least element of $J_{\bar{\rho}}$ such that every element of $J_{\bar{\rho}}$ is Σ_1-definable from parameters in $\bar{\alpha} \cup \{\bar{q}\}$ in $\langle J_{\bar{\rho}}, \bar{A}\rangle$.

Proof: Let $x \in J_{\bar{\rho}}$. Then $\sigma(x) \in J_\rho$, so for some $\vec{\delta} \in \alpha$, $\sigma(x)$ is Σ_1-definable from $q, \vec{\delta}$ in $\langle J_\rho, A\rangle$. Set $s = \langle\vec{\delta}\rangle$. Let ϕ be a Σ_0-formula such that:

(i) $\langle J_\rho, A\rangle \models \forall z\, \exists y\, \forall y'\, [y' = y \leftrightarrow \exists u \phi(u, y', z, q)]$;

(ii) $\langle J_\rho, A\rangle \models \forall z\, \forall y [\exists u \phi(u, y, z, q) \rightarrow (\exists \vec{\xi})\ (z = \langle\vec{\xi}\rangle)]$;

(iii) $(\forall y \in J_\rho)\ [y = \sigma(x) \leftrightarrow \langle J_\rho, A\rangle \models \exists u \phi(u, y, s, q)]$.

Let $<^*$ be the lexicographic ordering on $L \times L$ induced by $<_J$. Clearly $<^*$ is Δ_1^L and uniformly $\Delta_1^{J_\rho}$ for all limit $\rho > 0$. Let $\langle t, u_0\rangle$ be the $<^*$-least pair such that

$$\langle J_\rho, A\rangle \models \phi(u_0, \sigma(x), t, q).$$

Then $\langle t, u_0\rangle$ is Σ_1-definable from $\sigma(x), q$ in $\langle J_\rho, A\rangle$. So as $\sigma(x), q \in \mathrm{ran}(\sigma) \prec_1 \langle J_\rho, A\rangle, t, u_0 \in \mathrm{ran}(\sigma)$. Since $t \leq_J s$, $t \in J_\alpha$. Thus $t = \langle\vec{\xi}\rangle$ for some $\vec{\xi} \in \alpha$. By (i) above,

$$(\forall y \in J_\rho)\ [y = \sigma(x) \leftrightarrow \langle J_\rho, A\rangle \models \exists u \phi(u, y, t, q)].$$

Applying σ^{-1}, and setting $\bar{t} = \sigma^{-1}(t)$, we have

$$(\forall y \in J_{\bar{\rho}})\ [y = x \leftrightarrow \langle J_{\bar{\rho}}, \bar{A}\rangle = \exists u \phi(u, y, \bar{t}, \bar{q})].$$

Hence x is Σ_1-definable from parameters in $\bar{\alpha} \cup \{\bar{q}\}$ in $\langle J_{\bar{\rho}}, \bar{A}\rangle$. The minimality of $\bar{q}$ is proved just as in Claim C. □

Claim J: $\bar{\nu}$ is not Σ_n-regular over $J_{\bar{\beta}}$.

Proof: By Claim I,

$$J_{\bar{\rho}} = h_{\bar{\rho},\bar{A}} "(\omega \times (J_{\bar{\alpha}} \times \{\bar{q}\})).$$

In particular, there is a $\Sigma_1(\langle J_{\bar{\rho}},\bar{A}\rangle)$ map from a subset of $\bar{\alpha}$ onto $\bar{\nu}$. Since $\bar{\rho} = \rho_{\bar{\beta}}^{n-1}$, $\bar{A} = A_{\bar{\beta}}^{n-1}$, this map is $\Sigma_n(J_{\bar{\beta}})$. □

Claim K: $\bar{\beta} = \beta(\bar{\nu})$, $n = n(\bar{\nu})$, $\bar{\nu} \in R$, $\bar{\rho} = \rho(\bar{\nu})$, $\bar{A} = A(\bar{\nu})$.

Proof: By claims H and J. (For $\bar{\nu} \in R$, notice that $\tilde{\sigma}: J_{\bar{\beta}} \prec_1 J_\beta$, so $\lim(\beta) \to \lim(\bar{\beta})$.) □

Claim L: $\bar{q} = q(\bar{\nu})$ and $\sigma(p(\bar{\nu})) = p(\nu)$.

Proof: By claims I and K, $\bar{q} = q(\bar{\nu})$. If $\nu < \rho$, then $\bar{\nu} < \bar{\rho}$ and we have $p(\bar{\nu}) = \langle q(\bar{\nu}),\bar{\nu}\alpha_{\bar{\nu}}\rangle$, so $\sigma(p(\bar{\nu})) = p(\nu)$ by claims G and the fact that $\sigma(\bar{\nu}) = \nu$. If $\nu = \rho$, then $\bar{\nu} = \bar{\rho}$ and $p(\bar{\nu}) = \langle q(\bar{\nu}),\alpha_{\bar{\nu}}\rangle$, so again we get $\sigma(p(\bar{\nu})) = p(\nu)$. □

That completes the proof of lemma 2.6. □

Lemma 2.7

Let $\bar{\nu} \in S_{\bar{\alpha}}$, $\nu \in S_\alpha$, $\bar{\alpha} < \alpha$, and suppose $\bar{\nu}$ is a limit point of $S_{\bar{\alpha}}$. Let

$$\sigma: \langle J_{\rho(\bar{\nu})},A(\bar{\nu})\rangle \prec_1 \langle J_{\rho(\nu)},A(\nu)\rangle$$

be such that $\sigma(p(\bar{\nu})) = p(\nu)$.

Let $\nu' = \sup \sigma[\bar{\nu}]$. Then $\nu' \in S_\alpha$ and there is an embedding

$$\sigma': \langle J_{\rho(\bar{\nu})},A(\bar{\nu})\rangle \prec_1 \langle J_{\rho(\nu')},A(\nu')\rangle$$

such that $\sigma \restriction J_{\bar{\nu}} \subseteq \sigma'$ and $\sigma'(p(\bar{\nu})) = p(\nu')$.

Proof: Set $\beta = \beta(\nu)$, $n = n(\nu)$, $\rho = \rho(\nu)$, $A = A(\nu)$, $q = q(\nu)$, $p = p(\nu)$, $\bar{\rho} = \rho(\bar{\nu})$, $\bar{A} = A(\bar{\nu})$, $\bar{q} = q(\bar{\nu})$, $\bar{p} = p(\bar{\nu})$.

Case 1. $\nu \in P$. Set $\gamma = \gamma(\nu)$.

For each $m \in \omega$, set

$$X_m = \{x \in J_\gamma \mid x \text{ is } \Sigma_{m+1}\text{-definable from parameters in } \alpha \cup \{q\} \text{ in } J_\gamma\}$$

Thus $X_m \prec_m J_\gamma$. Since α is the largest cardinal in J_ν and $\alpha \subseteq X_m \prec_1 J_\gamma$, $X_m \cap \nu$ is transitive, so let $\nu_m = X_m \cap \nu$. There is, by its definition, a J_γ-definable map of α onto X_m, so as ν is regular over J_γ, $\nu_m < \nu$. But by choice of q, $\bigcup_{m<\omega} X_m = J_\gamma$, so $\sup_{m<\omega} \nu_m = \nu$. But for each m, ν_m is Σ_1-definable from p in J_ρ, so $\nu_m \in \mathrm{ran}(\sigma)$. Hence $\sigma[\bar{\nu}]$ is cofinal in ν. Thus $\nu' = \nu$ and the lemma is trivially valid.

Case 2. $\nu \in R$.
$S_{\bar{\alpha}} \cap \bar{\nu}$ is $\Sigma_1^{J_{\bar{\nu}}}(\{\bar{\alpha}\})$ and $S_\alpha \cap \nu$ is $\Sigma_1^{J_\nu}(\{\alpha\})$ by the same definition. So, as $(\sigma \restriction J_{\bar{\nu}}): J_{\bar{\nu}} \prec_1 J_\nu$ and $\sigma(\bar{\alpha}) = \alpha$, $\nu' = \sup(S_\alpha \cap \nu')$. Since S_α is closed in $\sup(S_\alpha)$, we have $\nu' \in S_\alpha$.
Let $\eta = \sup(\sigma[\bar{\rho}])$, $\tilde{A} = A \cap J_\eta$. Since $\mathrm{ran}(\sigma) \subseteq J_\eta$, p, $\alpha \in J_\eta$. By Σ_0-absoluteness,

$$\sigma: \langle J_{\bar{\rho}}, \bar{A}\rangle \prec_0 \langle J_\eta, A\rangle.$$

So, as σ is cofinal in η,

$$\sigma: \langle J_{\bar{\rho}}, \bar{A}\rangle \prec_1 \langle J_\eta, \tilde{A}\rangle.$$

Set $\quad X = h_{\eta,\tilde{A}}"(\omega \times (J_\alpha \times \{q\}))$.

Let $\quad \pi: \langle J_\gamma, B\rangle \cong \langle X, \tilde{A} \cap X\rangle$.

Thus $\quad \pi: \langle J_\gamma, B\rangle \prec_1 \langle J_\eta, \tilde{A}\rangle$.

Claim A: $\mathrm{ran}(\sigma) \subseteq X$.

Proof: Let $x \in \mathrm{ran}(\sigma)$. Then x is Σ_1-definable from parameters in $\alpha \cup \{q\}$ in $\langle J_\rho, A\rangle$. Let $\bar{x} = \sigma^{-1}(x)$. An argument as in claim I of lemma 2.6 shows that $\bar{x}$ is Σ_1-definable from parameters in $\bar{\alpha} \cup \{\bar{q}\}$ in $\langle J_{\bar{\rho}}, \bar{A}\rangle$. So for some $i \in \omega$, $\bar{z} \in J_{\bar{\alpha}}$, $\bar{x} = h_{\bar{\rho},\bar{A}}(i, \langle \bar{z}, \bar{q}\rangle)$. Applying

$\sigma: \langle J_{\bar\rho}, \bar A\rangle \prec_1 \langle J_\eta, \tilde A\rangle$ and setting $z = \sigma(\bar z)$, we get $x = h_{\eta,\tilde A}(i,\langle z,q\rangle)$. Hence $x \in X$. □

<u>Claim B:</u> $X \cap \nu = \nu'$.

Proof: Let $\xi \in X \cap \nu$. Then for some $z \in J_\alpha$ and some $i \in \omega$, $\xi = h_{\eta,\tilde A}(i,\langle z,q\rangle)$. Since $\lim(\eta)$ there is a $\tau < \eta$ with $\xi = h_{\tau,\tilde A\cap J_\tau}(i,\langle z,q\rangle)$. Since $\eta = \sup(\sigma[\bar\rho])$ we can pick τ here so that $\tau = \sigma(\bar\tau)$ for some $\bar\tau < \bar\rho$. Set

$$\theta = \sup\,[\bar\nu \cap h_{\tau,\tilde A\cap J_\tau}"(\omega \times (J_\alpha \times \{q\}))],$$

$$\bar\theta = \sup\,[\bar\nu \cap h_{\bar\tau,\bar A\cap J_{\bar\tau}}"(\omega \times (J_{\bar\alpha} \times \{\bar q\}))].$$

Now, $\tilde A \cap J_\tau = A \cap J_\eta \cap J_\tau = A \cap J_\tau$, so $h_{\tau,\tilde A\cap J_\tau} \in J_\rho$ by amenability. So as $\alpha < \nu$ and ν is regular in J_ρ, $\theta < \nu$. Similarly $\bar\theta < \bar\nu$. But clearly, $\sigma(\bar\theta) = \theta$. Hence

$$\xi < \theta = \sigma(\bar\theta) < \sup(\sigma[\bar\nu]) = \nu'.$$

Thus $X \cap \nu \subseteq \nu'$.

Now let $\xi \in \nu'$. For some $\bar\delta < \bar\nu$, $\xi < \delta = \sigma(\bar\delta)$. Since $\bar\delta < \bar\nu$ there is $\bar f \in J_{\bar\nu}$, $\bar f: \bar\alpha \xrightarrow{\text{onto}} \bar\delta$. Since $(\sigma\restriction J_{\bar\nu}): J_{\bar\nu} \prec_0 J_\nu$, $f = \sigma(\bar f) \in J_\nu$, and $f: \alpha \xrightarrow{\text{onto}} \delta$. But by claim A, $f \in X$. So as $\alpha \subseteq X$, $\delta = f[\alpha] \subseteq X$. Hence $\xi \in X$. Thus $\nu' \subseteq X \cap \nu$. □

Now, $\operatorname{ran}(\sigma) \prec_1 \langle J_\nu, \tilde A\rangle$, $X \prec_1 \langle J_\eta, \tilde A\rangle$, and $\operatorname{ran}(\sigma) \subseteq X$. Hence $\operatorname{ran}(\sigma) \prec_1 \langle X, \tilde A \cap X\rangle$. So, setting $\sigma' = \pi^{-1} \circ \sigma$, we have

$$\sigma': \langle J_{\bar\rho}, \bar A\rangle \prec_1 \langle J_\gamma, B\rangle.$$

By claim B, $\pi^{-1}\restriction\nu' = \mathrm{id}\restriction\nu'$, so $\sigma'\restriction J_{\bar\nu} = \mathrm{id}\restriction J_{\bar\nu}$, giving $\sigma\restriction J_{\bar\nu} \subseteq \sigma'$. It suffices now to prove that $\gamma = \rho(\nu')$, $B = A(\nu')$, and $\pi^{-1}(p) = p(\nu')$. By Σ_0-absoluteness,

$$\pi: \langle J_\gamma, B\rangle \prec_0 \langle J_\rho, A\rangle.$$

So by the fine structure theory there is a unique β' such that $\gamma = \rho_{\beta'}^{n-1}$, $B = A_{\beta'}^{n-1}$, and a mapping $\tilde{\pi}: J_{\beta'} \prec_n J_\beta$ such that $\pi \subseteq \tilde{\pi}$. Set $q' = \pi^{-1}(q)$, $p' = \pi^{-1}(p)$.

Suppose $\nu < \rho$. Then $\bar{\nu} < \bar{\rho}$ and $p = \langle q, \nu\ \alpha \rangle$, $\bar{p} = \langle \bar{q}, \bar{\nu}, \bar{\alpha} \rangle$. Hence $p' = \langle q', \pi^{-1}(\nu), \pi^{-1}(\alpha) \rangle$. But by claim B, $\pi^{-1}(\nu) = \nu'$. Hence as $\pi^{-1}(\alpha) = \alpha$, $p' = \langle q', \nu', \alpha \rangle$. Moreover, $\nu' < \gamma$.

Suppose $\nu = \rho$. Then $\bar{\nu} = \bar{\rho}$ and $p = \langle q, \alpha \rangle$, $\bar{p} = \langle \bar{q}, \bar{\alpha} \rangle$. Hence $p' = \langle q', \alpha \rangle$. And, using claim B, $\nu' = \gamma$.

By the above, in order to show that $\pi^{-1}(p) = p(\nu')$ it suffices to show that $\pi^{-1}(q) = q(\nu')$, i.e. that $q' = q(\nu')$.

<u>Claim C:</u> ν' is Σ_{n-1}-regular over $J_{\beta'}$.

Proof: Just as in Claim H in lemma 2.6. □

<u>Claim D:</u> ν' is not Σ_n-regular over $J_{\beta'}$.

Proof: By claim B, $\nu' = \nu' \cap h_{\eta,\tilde{A}}$ "$(\omega \times (J_\alpha \times \{q\}))$.
So, as $\pi: \langle J_\gamma, B \rangle \prec_1 \langle J_\nu, \tilde{A} \rangle$ and $\nu' \subseteq X = \mathrm{ran}(\pi)$, we have $\nu' = \nu' \cap h_{\gamma,B}$ "$(\omega \times (J_\alpha \times \{q'\}))$. Thus there is a $\Sigma_1(\langle J_\gamma, B \rangle)$ map of α onto ν'. But $\gamma = \rho_{\beta'}^{n-1}$, $B = A_{\beta'}^{n-1}$, so this map is $\Sigma_n(J_{\beta'})$. □

<u>Claim E:</u> $\beta' = \beta(\nu')$, $n = n(\nu')$, $\nu' \in R$, $\gamma = \rho(\nu')$, $B = A(\nu')$.

Proof: By claims C and D. □

<u>Claim F:</u> $q' = q(\nu')$.

Proof: By definition, $X = h_{\eta,\tilde{A}}$ "$(\omega \times (J_\alpha \times \{q\}))$. So, applying π^{-1}, $J_\gamma = h_{\gamma,B}$ "$(\omega \times (J_\alpha \times \{q'\}))$. Hence every member of J_γ is Σ_1-definable from parameters in $\alpha \cup \{q'\}$ in $\langle J_\gamma, B \rangle$. And an argument as in Claim C in lemma 2.6 completes the proof. □

The lemma is proved. □

Suppose now that ν is a limit ordinal and $X \subseteq J_\nu$. We write $X \prec_Q J_\nu$ iff for all Σ_0-formulas $\phi(v_0,v_1)$ of set theory, with parameters from X,

$$X \models (\forall\alpha)\ (\exists\beta > \alpha)\ \ \phi(\beta,J_\beta) \quad \text{iff} \quad J_\nu \models (\forall\alpha)\ (\exists\beta > \alpha)\ \ \phi(\beta,J_\beta).$$

Clearly, if $X \prec_Q J_\nu$, then $X \prec_1 J_\nu$. Conversely, if $X \prec_1 J_\nu$ and $X \cap \nu$ is unbounded in ν, then $X \prec_Q J_\nu$.

We write $\pi\colon J_{\bar\nu} \prec_Q J_\nu$ iff $\mathrm{ran}(\pi) \prec_Q J_\nu$.

Define a binary relation $\dashrightarrow$ on S by $\nu \dashrightarrow \tau$ iff $\alpha_\nu < \alpha_\tau$ and there is an embedding $\sigma\colon \langle J_{\rho(\nu)},A(\nu)\rangle \prec_1 \langle J_{\rho(\tau)},A(\tau)\rangle$ such that

$\sigma\restriction\alpha_\nu = \mathrm{id}\restriction\alpha_\nu$, $\sigma(p(\nu)) = p(\tau)$, and $(\sigma\restriction J_\nu)\colon J_\nu \prec_Q J_\tau$.

Clearly, $\dashrightarrow$ is a well-founded partial ordering on S. We show that $\dashrightarrow$ is a tree. It suffices to show that if $\tau \in S_\alpha$ and $\bar\alpha < \alpha$, there is at most one $\nu \dashrightarrow \tau$ with $\alpha_\nu = \bar\alpha$. Let $\nu \dashrightarrow \tau$, $\alpha_\nu = \bar\alpha$. Then there is a Σ_1-embedding $\sigma\colon \langle J_{\rho(\nu)},A(\nu)\rangle \prec_1 \langle J_{\rho(\tau)},A(\tau)\rangle$ as above. Now,
$J_{\rho(\nu)} = h_{\rho(\nu),A(\nu)}$ "$(\omega \times (J_{\bar\alpha} \times \{p(\nu)\}))$, so
$\mathrm{ran}(\sigma) = h_{\rho(\tau),A(\tau)}$ "$(\omega \times (J_{\bar\alpha} \times \{p(\tau)\}))$. Thus $\mathrm{ran}(\sigma)$ is entirely determined by τ and $\bar\alpha$. Hence $\rho(\nu)$ is completely determined by τ and $\bar\alpha$. Hence so is ν.

By lemma 2.5, if $\nu \dashrightarrow \tau$, the map σ which testifies this fact is unique, so we may denote if by $\sigma_{\nu\tau}$. Set $\pi_{\nu\tau} = (\sigma_{\nu\tau}\restriction\nu) \cup \{\langle\tau,\nu\rangle\}$. The systems $\{\sigma_{\nu\tau} \mid \nu \dashrightarrow \tau\}$ and $\{\pi_{\nu\tau} \mid \nu \dashrightarrow \tau\}$ are commutative.

<u>Lemma 2.8</u>

Let $\nu \dashrightarrow \tau$. Then $\pi_{\nu\tau}$ maps $S_{\alpha_\nu} \cap (\nu+1)$ into $S_{\alpha_\tau} \cap (\tau+1)$ in an order-preserving fashion such that:

(i) if $\gamma = \min(S_{\alpha_\nu})$, then $\pi_{\nu\tau}(\gamma) = \min(S_{\alpha_\tau})$;

(ii) if γ immediately succeeds δ in $S_{\alpha_\nu} \cap (\nu+1)$, then $\pi_{\nu\tau}(\gamma)$ immediately succeeds $\pi_{\nu\tau}(\delta)$ in $S_{\alpha_\tau} \cap (\tau+1)$;

(iii) if γ is a limit point of $S_{\alpha_\nu} \cap (\nu+1)$, then $\pi_{\nu\tau}(\nu)$ is a limit point of $S_{\alpha_\tau} \cap (\tau+1)$.

Proof: This follows trivially from lemma 2.1 (vi), except in the following case. Suppose ν is a limit point of S_{α_ν} and that $\nu = \rho(\nu)$. We wish to show that $\tau = \pi_{\nu\tau}(\nu)$ is a limit point of S_{α_τ}. This follows easily from the fact that $(\sigma_{\nu\tau}\restriction J_\nu): J_\nu \prec_Q J_\tau$. This is the only point where the Q-embedding condition is required. □

Lemma 2.9

Let $\bar{\tau} \dashv \tau$, $\bar{\nu} \in S_{\alpha_{\bar{\tau}}} \cap \bar{\tau}$, $\nu = \pi_{\bar{\tau}\tau}(\bar{\nu})$. Then $\bar{\nu} \dashv \nu$, $\pi_{\bar{\nu}\nu}\restriction\bar{\nu} = \pi_{\bar{\tau}\tau}\restriction\bar{\nu}$, and $\sigma_{\bar{\nu}\nu} = \sigma_{\bar{\tau}\tau}\restriction J_{\rho(\bar{\nu})}$.

Proof: By lemma 2.4, $\sigma_{\bar{\tau}\tau}(\rho(\bar{\nu})) = \rho(\nu)$, $\sigma_{\bar{\tau}\tau}(A(\bar{\nu})) = A(\nu)$, and $\sigma_{\bar{\tau}\tau}(p(\bar{\nu})) = p(\nu)$. The lemma follows immediately. □

Lemma 2.10

If $\tau \in S$ is a limit point of $\dashv$, then $\tau = \bigcup_{\nu\dashv\tau} \pi_{\nu\tau}[\nu]$ and $J_{\rho(\tau)} = \bigcup_{\nu\dashv\tau} \sigma_{\nu\tau}[J_{\rho(\nu)}]$.

Proof: We commence by proving that $\alpha_\tau = \sup\{\alpha_\nu | \nu \dashv \tau\}$. Suppose not, and let $\bar{\alpha} = \sup\{\alpha_\nu | \nu \dashv \tau\} < \alpha_\tau$. For each $\nu \dashv \tau$, let $X_\nu = \mathrm{ran}(\sigma_{\nu\tau})$. Thus $\langle X_\nu | \nu \dashv \tau\rangle$ is an increasing chain of Σ_1 submodels of $\langle J_{\rho(\tau)}, A(\tau)\rangle$. Set $X = \bigcup\{X_\nu | \nu \dashv \tau\}$. Thus $X \prec_1 \langle J_{\rho(\tau)}, A(\tau)\rangle$. Let

$$\sigma: \langle J_{\bar{\rho}}, \bar{A}\rangle \cong \langle X, A(\tau) \cap X\rangle.$$

By lemma 2.6 there is a unique $\bar{\nu}$ such that $\bar{\rho} = \rho(\bar{\nu})$, $\bar{A} = A(\bar{\nu})$, and $\sigma(p(\bar{\nu})) = p(\tau)$. Since $\bar{\alpha} = \sup\{\alpha_\nu | \nu \dashv \tau\}$, we clearly have $\bar{\nu} \in S_{\bar{\alpha}}$. And since $X_\nu \cap J_\tau \prec_Q J_\tau$ for all $\nu \dashv \tau$, we have $X \cap J_\tau \prec_Q J_\tau$. Hence $\bar{\nu} \dashv \tau$ and $\sigma = \sigma_{\bar{\nu}\tau}$. Thus τ succeeds $\bar{\nu}$ in $\dashv$, contrary to the assumption on τ.

Now let $x \in J_{\rho(\tau)}$. Then for some $\vec{\delta} \in \alpha_\tau$, x is Σ_1-definable from $p(\tau)$, $\vec{\delta}$ in $\langle J_{\rho(\tau)}, A(\tau)\rangle$. Pick $\nu \dashv \tau$ with $\vec{\delta} \in \alpha_\nu$ (by the above). Then, since $\vec{\delta}, p(\tau) \in \mathrm{ran}(\sigma_{\nu\tau}) \prec_1 \langle J_{\rho(\tau)}, A(\tau)\rangle$, we have $x \in \mathrm{ran}(\sigma_{\nu\tau})$. This shows that $J_{\rho(\tau)} = \bigcup_{\nu \dashv \tau} \sigma_{\nu\tau}[J_{\rho(\nu)}]$. The corresponding result for π follows at once. □

Lemma 2.11

If τ is not maximal in S_α, then τ is a limit point of $\dashv$.

Proof: Let $\alpha = \alpha_\tau$, and pick $\lambda \in S_\alpha$, $\lambda > \tau$. Given $\theta < \alpha$ we show that there is $\nu \dashv \tau$ such that $\alpha_\nu \geq \theta$. We place ourselves in J_λ. Let X be the smallest elementary submodel of $\langle J_{\rho(\tau)}, A(\tau)\rangle$ containing $p(\tau)$ and θ, such that $X \cap \alpha$ is transitive. Since $\alpha = \omega_1$, $X \cap \alpha \in \alpha$. Let $\bar{\alpha} = X \cap \alpha$. Now return to the real world. Let $\sigma: \langle J_\rho, A\rangle \cong \langle X, A(\tau) \cap X\rangle$. By lemma 2.6 there is a unique $\nu \in S_{\bar{\alpha}}$ such that $\rho = \rho(\nu)$, $A = A(\nu)$. Then $\nu \dashv \tau$ and $\sigma_{\nu\tau} = \sigma$. Since $\alpha_\nu = \bar{\alpha} > \theta$, we are done. □

Lemma 2.12

Let $\bar{\nu} \dashv \nu$, and suppose $\bar{\nu}$ is a limit point of $S_{\alpha_{\bar{\nu}}}$. Let $\nu' = \sup(\pi_{\bar{\nu}\nu}[\bar{\nu}])$. Then $\bar{\nu} \dashv \nu'$ and $\sigma_{\bar{\nu}\nu'} \upharpoonright J_{\bar{\nu}} = \sigma_{\bar{\nu}\nu} \upharpoonright J_{\bar{\nu}}$.

Proof: Clearly, $(\sigma_{\bar{\nu}\nu} \upharpoonright J_{\bar{\nu}}): J_{\bar{\nu}} \prec_0 J_{\nu'}$. But $\sigma_{\bar{\nu}\nu}[\bar{\nu}]$ is cofinal in ν'. Hence $(\sigma_{\bar{\nu}\nu} \upharpoonright J_{\bar{\nu}}): J_{\bar{\nu}} \prec_Q J_{\nu'}$. The lemma follows easily from lemma 2.7 now. □

§ 3. A New Construction of a Souslin $\aleph_2$-Tree

We use the above theory to construct a Souslin $\aleph_2$-tree in L. We assume V = L throughout.

We define, by recursion on τ, trees T^τ, $\tau \in S$. T^τ will be a normal tree of height $\mathrm{otp}(S_{\alpha_\tau} \cap \tau)$, and we shall use the members of $S_{\alpha_\tau} \cap \tau$ to index the levels of T^τ (so $T^\tau = \bigcup\{T^\tau_\nu \mid \nu \in S_{\alpha_\tau} \cap \tau\}$). If $\tau < \omega_1$, T^τ will have width $\aleph_1$, and if $\tau \in S_{\omega_1}$, T^τ will have width $\aleph_2$. Our final $\aleph_2$-tree, T, will be a subtree of the tree $\bigcup\{T^\tau \mid \tau \in S_{\omega_1}\}$.

For $\nu \in S$, we denote by $s(\nu)$ the immediate successor of ν in S_{α_ν}, if $\nu \neq \max(S_{\alpha_\nu})$, with $s(\nu)$ undefined if $\nu = \max(S_{\alpha_\nu})$.

We carry out the construction so as to preserve the following conditions:

(P1) If $\nu \in S_{\alpha_\tau} \cap \tau$, T^τ will be an end-extension of T^ν (i.e. we shall have $T^\nu = \bigcup\{T^\tau_\gamma \mid \gamma \in S_{\alpha_\tau} \cap \nu\}$);

(P2) $T^\tau_\gamma = s(\gamma) - \gamma$;

(P3) If $\bar\tau \dashv \tau$, $\pi_{\bar\tau\tau} \restriction \bar\tau$ embeds $T^{\bar\tau}$ into T^τ so that whenever $\bar\nu \in S_{\alpha_{\bar\tau}} \cap \bar\tau$ and $\nu = \pi_{\bar\tau\tau}(\bar\nu)$, then $\pi_{\bar\tau\tau}[T^{\bar\tau}_{\bar\nu}] \subseteq T^\tau_\nu$.

The definition of T^τ falls into three cases, depending upon the position of τ in $\dashv$. We consider these cases in order of increasing complexity.

<u>Case 1.</u> τ is a limit point in $\dashv$.

In this case we set $T^\tau = \bigcup_{\nu \dashv \tau} \pi_{\nu\tau}[T^\nu]$. By lemma 2.10, this defines a normal tree of height $\mathrm{otp}(S_{\alpha_\tau} \cap \tau)$ on τ such that $\pi_{\nu\tau} \restriction \nu$ embeds T^ν into T^τ (to satisfy (P3)) for each $\nu \dashv \tau$. We must check that if $\eta \in S_{\alpha_\tau} \cap \tau$, then T^τ is an end-extension of T^η. But by lemma 2.11, η is a limit point of $\dashv$, and by lemma 2.10, if $\nu \dashv \tau$ is sufficiently large,

$\eta = \pi_{\nu\tau}(\bar{\eta})$ for some $\bar{\eta} \in S_{\alpha_\nu} \cap \nu$, whence by lemma 2.9, $\sigma_{\bar{\eta}\eta} \restriction T^{\bar{\eta}} = \sigma_{\nu\tau} \restriction T^{\bar{\eta}}$, and the result follows. There is nothing further to check.

Case 2. τ is minimal in $\triangleleft$.

There are three subcases to consider.

Case 2.1 τ is initial in S_{α_τ}.

Then we must set $T^0 = \phi$. There is nothing for us to check.

Case 2.2 τ is a limit point of S_{α_τ}.

Set $T^\tau = \cup\{T^\nu | \nu \in S_{\alpha_\tau} \cap \tau\}$. There is nothing to check.

Case 2.3 $\tau = s(\nu)$.

There are three subcases to consider.

Case 2.3.1 ν is initial in S_{α_τ}.

Set $T^\tau_\nu = \tau - \nu$. There is nothing to check.

Case 2.3.2 $\nu = s(\eta)$.

Using the ordinals of $\tau - \nu$, appoint infinitely many extensions of each member of T^ν_η to form T^τ_ν, and set $T^\tau = T^\nu \cup T^\tau_\nu$. There is nothing to check.

Case 2.3.3 ν is a limit point of S_{α_τ}.

First note that by lemma 2.11, τ is maximal in S_{α_τ}, so by 2.1 (iii), $\tau \notin S_{\omega_1}$. Hence ν is a countable limit ordinal. Hence $\text{otp}(S_{\alpha_\tau} \cap \nu)$ is a countable limit ordinal, and for each $x \in T^\nu$ we can pick a branch b_x through T^ν, containing x. Using the ordinals in $\tau - \nu$, we appoint one-point extensions of each branch b_x to obtain T^τ_ν, whence $T^\tau = T^\nu \cup T^\tau_\nu$. The actual choice of the branches b_x is not important except in the following case. Suppose $T^\nu \in J_{\rho(\nu)}$. (Notice that we make extensive use of the countability of the ordinals in $S \cap \omega_1$ in order to construct the trees T^τ, $\tau \in S \cap \omega_1$, so it will only be "accidental" that T^ν lies in $J_{\rho(\nu)}$. However, it certainly will occur, as we show later.) In this case, let Q be the set of all thin initial segments of T^ν which lie in $J_{\rho(\nu)}$. Q is countable, so it is easy to pick the branches b_x to be eventually disjoint from each member of Q.

That completes the definition in this case. There is nothing for us to check.

<u>Case 3.</u> τ immediately succeeds $\bar{\tau}$ in $\dashv$.

As in Case 2, there are three subcases.

<u>Case 3.1</u> τ is initial in S_{α_τ}.

Set $T^\tau = \phi$. There is nothing to check.

<u>Case 3.2</u> τ is a limit point of S_{α_τ}.

Set $T^\tau = \cup\{T^\nu | \nu \in S_{\alpha_\tau} \cap \tau\}$. The only thing to check is that $\pi_{\bar{\tau}\tau} \upharpoonright \bar{\tau}$ embeds $T^{\bar{\tau}}$ into T^τ so that whenever $\bar{\nu} \in S_{\alpha_\tau} \cap \bar{\tau}$ and $\nu = \pi_{\bar{\tau}\tau}(\bar{\nu})$, then $\pi_{\bar{\tau}\tau}[T^{\bar{\tau}}_{\bar{\nu}}] \subseteq T^\tau_\nu$. Now, by lemma 2.8, $\bar{\tau}$ will be a limit point of $S_{\alpha_{\bar{\tau}}}$, so $\pi_{\bar{\tau}\tau} \upharpoonright \bar{\tau} = \cup\{\pi_{\bar{\tau}\tau} \upharpoonright \bar{\nu} \mid \bar{\nu} \in S_{\alpha_\tau} \cap \bar{\tau}\}$. The result follows immediately using Lemma 2.9.

<u>Case 3.3</u> $\tau = s(\nu)$.

Then $\bar{\tau} = s(\bar{\nu})$, where $\pi_{\bar{\tau}\tau}(\bar{\nu}) = \nu$. There are three subcases to consider.

<u>Case 3.3.1</u> ν is initial in S_{α_τ}.

Let $T^\tau_\nu = \tau - \nu$. There are no non-trivial checks to be made.

<u>Case 3.3.2</u> $\nu = s(\eta)$.

Then $\bar{\nu} = s(\bar{\eta})$, where $\pi_{\bar{\tau}\tau}(\bar{\eta}) = \eta$. For each pair $x, y \in T^{\bar{\tau}}$ such that $x \in T^{\bar{\tau}}_{\bar{\eta}}$, $y \in T^{\bar{\tau}}_{\bar{\nu}}$, and $x <_T y$, let $\pi_{\bar{\tau}\tau}(y)$ extend $\pi_{\bar{\tau}\tau}(x)$ in T^τ_ν. Now, $\tau - (\nu \cup \pi_{\bar{\tau}\tau}[T^{\bar{\tau}}_{\bar{\nu}}])$ is infinite (this is clear from the construction of section 2). Use the ordinals in this set to ensure that every element of T^τ_η has infinitely many extensions in T^τ_ν. There is nothing of a non-trivial nature which needs to be checked.

<u>Case 3.3.3</u> ν is a limit point of S_{α_τ}.

There are two subcases to consider.

<u>Case 3.3.3.1</u> $\pi_{\bar{\tau}\tau}[\bar{\nu}]$ is cofinal in ν.

For each $x \in T^{\bar{\tau}}_{\bar{\nu}}$, let $\pi_{\bar{\tau}\tau}(x)$ be an extension in $T^{\tau}{}_{\nu}$ of the branch through T^{ν} determined by $\{\pi_{\bar{\tau}\tau}(y) | y <_T x\}$. Using the ordinals in the infinite set $\tau - (\nu \cup \pi_{\bar{\tau}\tau}[T^{\bar{\tau}}_{\bar{\nu}}])$, we now complete the definition of $T^{\tau}{}_{\nu}$ by extending two further collections of branches through T^{ν}.

Firstly, if $\gamma \dashv \nu$ and $\gamma \neq \max(S_{\alpha_\gamma})$ and $\pi_{\gamma\nu}[\gamma]$ is cofinal in ν, and if $x \in T^{s(\gamma)}_{\gamma}$, we ensure that the branch $\{\pi_{\gamma\nu}(y) | y <_T x\}$^though T^{ν} extends onto $T^{\tau}{}_{\nu}$.

Secondly, for each $x \in T^{\nu}$ we pick a branch b_x through T^{ν}, containing x, and distinct from each of the branches extended above, and extend that branch onto $T^{\tau}{}_{\nu}$. The choice of the b_x is made according to the rules laid out in case 2.3.3.

There are no non-trivial checks to be made.

<u>Case 3.3.3.2</u> $\lambda = \sup \pi_{\bar{\nu}\nu}[\bar{\nu}] < \nu$.

By lemma 2.12, $\lambda \in S_{\alpha_\nu}$, $\bar{\nu} \dashv \lambda$, and $\sigma_{\bar{\nu}\lambda} \restriction J_{\bar{\nu}} = \sigma_{\bar{\nu}\nu} \restriction J_{\bar{\nu}}$. Moreover, (using lemma 2.9), $\pi_{\bar{\nu}\lambda} \restriction \bar{\nu} = \pi_{\bar{\tau}\tau} \restriction \bar{\nu}$. Let $\eta = s(\lambda)$. Choose $\eta_o \dashv \eta$ with $\alpha_{\eta_o} > \alpha_{\bar{\nu}}$. Let $\eta_1 \dashv \eta$ be an immediate successor of η_o in $\dashv$. Let $\eta_o = s(\lambda_o)$, $\eta_1 = s(\lambda_1)$. Each of λ_o, λ_1 is a limit point of its level. Moreover, $\pi_{\bar{\nu}\lambda} = \pi_{\lambda_1\lambda} \cdot \pi_{\lambda_o\lambda_1} \cdot \pi_{\bar{\nu}\lambda_o}$, so in particular $\pi_{\lambda_o\lambda_1}[\lambda_o]$ is cofinal in λ_1. Hence Case 3.3.3.1 applied in the definition of T^{η_1}.

Let $x \in T^{\bar{\tau}}_{\bar{\nu}}$. By the construction in Case 3.3.3.1 for T^{η_1}, we know that there is a point $y(x)$ in $T^{\eta_1}_{\lambda_1}$ which extends all the points $\pi_{\bar{\nu}\lambda_1}(z)$

for $z <_T x$. Let $y'(x) = \pi_{\eta_1\eta}(y(x))$. Then $y'(x) \in T^{\eta}{}_{\lambda}$ extends all the points $\pi_{\bar{\nu}\nu}(z)$ $(= \pi_{\bar{\nu}\lambda}(z) = \pi_{\lambda_1\lambda} \cdot \pi_{\bar{\nu}\lambda}(z))$ for $z <_T x$. Pick a branch d_x through T^{ν}, containing $y'(x)$, and let $\pi_{\bar{\tau}\tau}(x)$ extend d_x on $T^{\tau}{}_{\nu}$. Also, for each $x \in T^{\nu}$, pick a branch b_x through T^{ν}, containing x, and distinct from the branches d_x just extended, and use the ordinals in $\tau - (\nu \cup \pi_{\bar{\tau}\tau}[T^{\bar{\tau}}{}_{\bar{\nu}}])$ to extend these branches onto $T^{\tau}{}_{\nu}$.
The choice of the branches d_x, $x \in T^{\bar{\tau}}{}_{\bar{\nu}}$, and b_x, $x \in T^{\nu}$, is made as in Case 2.3.3.

There are no non-trivial checks to be made.

That completes the construction. Let $T = \cup\{T^{\tau} \mid \tau \in S_{\omega_1}\}$. T is clearly an $\aleph_2$-tree.

<u>Lemma 3.1</u>
T has at most countably many ω_2-branches.

Proof: Suppose otherwise, and let $\{b_{\nu} \mid \nu < \omega_1\}$ be $\aleph_1$ many distinct ω_2-branches of T. Let $\delta < \omega_2$ be least such that $\nu < \tau < \omega_1 \rightarrow b_{\nu} \cap T_{\delta} \neq$ $\neq b_{\tau} \cap T_{\delta}$. Let $U = \bigcup_{\nu<\omega_1} b_{\nu}$. Clearly, U is a thin initial segment of T.

Let M be the smallest $M \prec J_{\omega_3}$ such that $\langle T,U,\delta\rangle \in M$ and $M \cap \omega_2$ is transitive. Let $\nu = M \cap \omega_2$. Clearly, $\nu \in \omega_2$.

Let $\pi: M \cong J_{\lambda}$. Then, $\pi \restriction \nu = \mathrm{id} \restriction \nu$, $\pi(\omega_2) = \nu$, $\pi(T) = T \cap \nu$, and $\pi(U) = U \cap \nu$. It is easily seen that ν is a limit point of S_{ω_1}, and that $T \cap \nu = T^{\nu}$, $U \cap \nu = U \cap T^{\nu}$. Moreover, $U \cap \nu$ is a thin initial segment of T^{ν}.

<u>Claim.</u> $\beta(\nu) = \lambda + 1$ and $n(\nu) = 1$. Moreover, $\mathrm{cf}(\nu) = \omega$.

Proof: Since $\pi^{-1}: J_{\lambda} \prec J_{\omega_3}$ and $\pi^{-1}(\nu) = \omega_2$, ν is clearly regular over J_{λ}. We show that ν is Σ_1-singular over $\lambda + 1$, and that $\mathrm{cf}(\nu) = \omega$.

Let X_o be the smallest $X \prec_1 J_\lambda$ such that $\langle T \cap \nu, U \cap \nu, \delta\rangle \in X$, and set $\nu_o = \sup(X_o \cap \nu)$. Notice that since X_o is definable in J_λ, and ν is regular over J_λ, we must have $\nu_o < \nu$.

Now suppose $X_n \prec_{n+1} J_\lambda$ and $\nu_n < \nu$ are defined. Let X_{n+1} be the smallest $X \prec_{n+2} J_\lambda$ such that $\langle T \cap \nu, U \cap \nu, \delta\rangle \in X$ and $\nu_n \subseteq X$, and set $\nu_{n+1} = \sup(X_{n+1} \cap \nu)$. Since X_{n+1} is J_λ-definable and ν is regular over J_λ and $\nu_n < \nu$, we have $\nu_{n+1} < \nu$.

Let $X = \bigcup_{n<\omega} X_n$. Clearly, $X \prec J_\lambda$ and $\langle T \cap \nu, U \cap \nu, \delta\rangle \in X$, and $X \cap \nu = \sup_{n<\omega} \nu_n$ (which is transitive, of course). But since $J_\lambda \simeq M$, J_λ is the smallest elementary submodel of J_λ which contains $\langle T \cap \nu, U \cap \nu, \delta\rangle$ and has a transitive intersection with ν. Thus $X = J_\lambda$. Thus, in particular $\sup_{n<\omega}\nu_n = \nu$. But clearly, $\langle X_n | n < \omega\rangle$ and $\langle \nu_n | n < \omega\rangle$ are Σ_1-definable over $J_{\lambda+1}$. This proves the claim. □

By the claim, we have, in particular, $\rho(\nu) = \beta(\nu) = \lambda + 1$. Hence as $T \cap \nu, U \cap \nu \in J_\lambda$, we have $T^\nu, U \cap \nu \in J_{\rho(\nu)}$.

Now, since $\nu > \delta$, there are $\aleph_1$ many points in $\dot{T}_\nu$ which extend a branch through U. We obtain our desired contradiction by demonstrating that our construction ensures that only countably many branches through $U \cap \nu$ extend onto T_ν.

By lemmas 2.1 (iii), 2.11, and 2.10, ν is a limit point in $\dashv$ and $J_{\rho(\nu)} = \bigcup_{\nu_o \dashv \nu} \sigma_{\nu_o\nu}[J_{\rho(\nu_o)}]$. So we can pick $\nu_o \dashv \nu$ large enough to have $T^\nu, U \cap \nu \in \sigma_{\nu_o\nu}[J_{\rho(\nu_o)}]$. Moreover, by our above claim, $cf(\nu) = \omega$, so we can pick ν_o here large enough for $\pi_{\nu_o\nu}[\nu_o]$ to be cofinal in ν. Notice that if $\nu_o \,\underline{\dashv}\, \nu_1 \dashv \nu$, then $\sigma_{\nu_1\nu}{}^{-1}(T^\nu) = T^{\nu_1}$ and $\sigma_{\nu_1\nu}{}^{-1}(U \cap \nu)$ is a thin initial segment of T^{ν_1} (because $\sigma_{\nu_1\nu}: J_{\rho(\nu_1)} \prec_1 J_{\rho(\nu)}$ and $\pi_{\nu_1\nu}[T^{\nu_1}] = T^\nu \cap \operatorname{ran}(\pi_{\nu_1\nu})$).

Let $C = \{\bar{\nu} \dashv \nu_o \mid \bar{\nu} \nmid \max(S_{\alpha_{\bar{\nu}}})\}$, and for $\bar{\nu} \in C$ let $B_{\bar{\nu}}$ be the set of all branches through $T^{\bar{\nu}}$ which extend on $T_{\bar{\nu}}^{s(\bar{\nu})}$. Let $B = \bigcup_{\bar{\nu} \in C} B_{\bar{\nu}}$. Notice that B is countable. We prove that if $\nu_o \dashv \nu_1 \dashv \nu$ and $\nu_1 \nmid \max(S_{\alpha_{\nu_1}})$, the only branches through $\sigma_{\nu_1\nu}{}^{-1}(U \cap \nu)$ which are extended on $T_{\nu_1}^{s(\nu_1)}$ are of the form $\pi_{\nu_o\nu_1}[b]^\frown$ for some $b \in B$. Since this is true in particular for $\nu_1 = \nu$, this will complete our proof of the lemma.

The proof is by induction on ν_1. Suppose the result holds for all $\nu_1' \dashv \nu_1$. Let $\tau_1 = s(\nu_1)$. There are three cases to consider.

<u>Case 1.</u> τ_1 is a limit point in $\dashv$.
Then $T^{\tau_1} = \bigcup_{\tau_2 \dashv \tau_1} \pi_{\tau_2\tau_1}[T^{\tau_2}]$, and the result follows easily from the induction hypothesis.

<u>Case 2.</u> τ_1 is minimal in $\dashv$.
In this case, <u>no</u> branch through $\sigma_{\nu_1\nu}{}^{-1}(U \cap \nu)$ is extended onto $T_{\nu_1}^{\tau_1}$.

<u>Case 3.</u> τ_1 succeeds τ_2 in $\dashv$.
Notice that τ_1 falls under Case 3.3.3.1 in the definition of T^{τ_1}. Suppose that $z \in T_{\nu_1}^{\tau_1}$ extends a branch through $\sigma_{\nu_1\nu}{}^{-1}(U \cap \nu)$. By our construction of $T_{\nu_1}^{\tau_1}$, the branch extended by z is not one of the branches b_x, since these are all eventually disjoint from $\sigma_{\nu_1\nu}{}^{-1}(U \cap \nu)$, which is a thin initial segment of T^{ν_1} in $J_{\rho(\nu_1)}$. There are two remaining possibilities.

Suppose first that $z = \pi_{\tau_2\tau_1}(x)$, where $x \in T_{\nu_2}^{\tau_2}$. Since $\sigma_{\nu_2\nu}{}^{-1}(U \cap \nu) = \sigma_{\nu_2\nu_1}{}^{-1} \circ \sigma_{\nu_1\nu}{}^{-1}(U \cap \nu)$, we see that x must extend a branch through $\sigma_{\nu_2\nu}{}^{-1}(U \cap \nu)$. So, by induction hypothesis, the branch extended by x is of the form $\pi_{\nu_o\nu_2}[b]^\frown$ for some $b \in B$. Hence the branch extended by z is of the form $\pi_{\nu_o\nu_1}[b]^\frown = \pi_{\nu_2\nu_1} \cdot \pi_{\nu_o\nu_1}[b]^\frown$ (for this b).

Now suppose that for some $\gamma \dashv \nu_1$ such that $\gamma \nmid \max(S_{\alpha_\gamma})$ and $\pi_{\gamma\nu_1}[\gamma]$ is cofinal in ν_1, z extends $\{\pi_{\gamma\nu_1}(y) | y <_T x\}^\frown$ where $x \in T_\gamma^{s(\gamma)}$. If $\gamma \underline{\dashv} \nu_o$, then $\gamma \in C$, so $\{y | y <_T x\} \in B$ and we are done. Otherwise, $\nu_o \dashv \gamma$, in which case $\{y | y <_T x\} \subseteq \sigma_{\gamma\nu}^{-1}(U \cap \nu)$ $(= \sigma_{\gamma\nu_1}^{-1} \cdot \sigma_{\nu_1\nu}(U \cap \nu))$, and by the induction hypothesis at γ, $\{y | y <_T x\} = \pi_{\nu_o\gamma}[b]^\frown$ for some $b \in B$, whence z extends $\pi_{\nu_o\nu_1}[b]^\frown$ $(= \gamma_{\gamma\nu_1} \cdot \pi_{\nu_o\nu_1}[b]^\frown)$, and again we are done. □

By lemma 3.1 we can find a point x of T such that no ω_2-branch of T contains x. We show that $\{y \in T | x <_T y\}$ is a Souslin $\aleph_2$-tree. In order to avoid carrying x along as a parameter, however, we shall simply assume that T already is Aronszajn. This clearly causes no loss of generality in our proof.

Lemma 3.2

T is Souslin.

Proof: Suppose not. Let X be an antichain of T of cardinality $\aleph_2$. Let

$$U = \{x \in \hat{X} | x \text{ has } \aleph_2 \text{ many successors in } \hat{X}\}.$$

It is easily seen that $|U| = \aleph_2$. Since T is Aronszajn, each member of U "splits" within U (i.e. for each $x \in U$ there are distinct $y, z \in U$ such that $x <_T y$, $x <_T z$, and $ht(y) = ht(z)$). Thus there is a $\delta < \omega_2$ such that $|U \cap T_\delta| = \aleph_1$. Arguing exactly as in lemma 3.1 now, we can find a $\nu > \delta$ such that $U \cap (T \restriction \nu)$ has at most countably many extensions in T_ν, contrary to the fact that each member of U has extensions in U on all higher levels of T. □

§ 4. A New Construction of a Kurepa $\aleph_2$-Tree

We need some preliminaries. A tree will be called almost normal if it has the property that each point has infinitely many extensions on all higher levels, but does not necessarily have the property that distinct points on a limit level have distinct sets of predecessors.

Lemma 4.1

Suppose T is an almost normal tree of height ω_2 and width $\aleph_2$, having $\aleph_3$ many ω_2-branches. Then there is a Kurepa $\aleph_2$-tree T' such that $\bigcup_{\alpha<\omega_2} T_{\alpha+1} = \bigcup_{\alpha<\omega_2} T'_{\alpha+1}$.

Proof: By discarding all limit levels of T, we obtain an almost normal tree T^o of height ω_2 and width $\aleph_2$ such that for every limit ordinal $\alpha < \omega_2$, every α-branch of $T^o \upharpoonright \alpha$ which extends on T^o_α has infinitely many extensions on T^o_α. Note that

$$T^o = \bigcup_{\alpha<\omega_2} T_{\alpha+1}.$$

For each limit ordinal $\alpha < \omega_2$ now and each α-branch, b, of $T^o \upharpoonright \alpha$ which is extended on T^o_α, introduce a new element which extends all the points of b and lies below all points of T^o_α which extend b. This defines an $\aleph_2$-tree T' such that $T^o = \bigcup_{\alpha<\omega_2} T'_{\alpha+1}$. Since T has $\aleph_3$ many ω_2-branches and $\bigcup_{\alpha<\omega_2} T'_{\alpha+1} = \bigcup_{\alpha<\omega_2} T_{\alpha+1}$, T' is clearly Kurepa. □

And so to the construction of a Kurepa tree (in L). What we in fact do is construct a tree T satisfying the requirements specified in lemma 4.1 above, and rely upon lemma 4.1 itself to extract the required Kurepa tree. We assume V = L from now on.

We construct trees T^τ, $\tau \in S$, in a manner similar to that employed in section 3. T^τ will be an almost normal tree of height $otp(S_{\alpha_\tau} \cap \tau)$

whose levels we shall index by the ordinals in $S_{\alpha_\tau} \cap \tau$. T^τ will have width $\aleph_1$ for $\tau \in S \cap \omega_1$ and width $\aleph_2$ for $\tau \in S_{\omega_1}$. The elements of T^τ_γ will be the ordinals in $s(\gamma)-\gamma$ (for $\gamma \in S_{\alpha_\tau} \cap \tau$). We shall carry out the construction so as to preserve the following conditions:

(i) If $\nu \in S_{\alpha_\tau} \cap \tau$, T^τ is an end-extension of T^ν.

(ii) If $\bar{\tau} \dashv \tau$, $\pi_{\bar{\tau}\tau} \upharpoonright \bar{\tau}$ embeds $T^{\bar{\tau}}$ into T^τ in such a way that whenever $\bar{\nu} \in S_{\alpha_{\bar{\tau}}} \cap \bar{\tau}$ and $\pi_{\bar{\tau}\tau}(\bar{\nu}) = \nu$, then $\pi_{\bar{\tau}\tau}[T^{\bar{\tau}}_{\bar{\nu}}] \subseteq T^\tau_\nu$.

(iii) Suppose $\tau = s(\nu)$, where ν is a limit point of S_{α_τ}, and that T^ν is Σ_1-definable over $\langle J_{\rho(\nu)}, A(\nu)\rangle$. Then every branch through T^ν which is Σ_1-definable over $\langle J_{\rho(\nu)}, A(\nu)\rangle$ extends onto T^τ_ν.

(iv) Suppose τ, ν are as in (iii), and $\bar{\tau} \dashv \tau$, $\bar{\tau} = s(\bar{\nu})$, $\pi_{\bar{\tau}\tau}(\bar{\nu}) = \nu$. Suppose further that T^ν is Σ_1-definable over $\langle J_{\rho(\nu)}, A(\nu)\rangle$. Let B be a branch through T^ν which is Σ_1-definable over $\langle J_{\rho(\nu)}, A(\nu)\rangle$. Since every element of $J_{\rho(\nu)}$ is Σ_1-definable from elements of $\alpha_\nu \cup \{p(\nu)\}$ in $\langle J_{\rho(\nu)}, A(\nu)\rangle$, there are Σ_1-formulas ϕ, ψ of $\mathcal{L}(A)$ and elements $\vec{x}, \vec{y} \in \alpha_\nu$ such that for all $u, v \in J_{\rho(\nu)}$,

$$u <_{T^\nu} v \quad \text{iff} \quad \langle J_{\rho(\nu)}, A(\nu)\rangle \models \phi(u, v, \vec{x}, p(\nu))$$

$$u \in B \quad \text{iff} \quad \langle J_{\rho(\nu)}, A(\nu)\rangle \models \psi(u, \vec{y}, p(\nu)).$$

Suppose $\vec{x}, \vec{y} \in \alpha_{\bar{\nu}}$. Since $\sigma_{\bar{\nu}\nu} : \langle J_{\rho(\bar{\nu})}, A(\bar{\nu})\rangle \prec_1 \langle J_{\rho(\nu)}, A(\nu)\rangle$ and $\sigma_{\bar{\nu}\nu} \upharpoonright \alpha_{\bar{\nu}} = \mathrm{id} \upharpoonright \alpha_{\bar{\nu}}$ and $\sigma_{\bar{\nu}\nu}(p(\bar{\nu})) = p(\nu)$ and $\sigma_{\bar{\nu}\nu}[T^{\bar{\nu}}] = T^\nu \cap \mathrm{ran}(\sigma_{\bar{\nu}\nu})$, we clearly have, for $u, v \in J_{\rho(\bar{\nu})}$,

$$u <_{T^{\bar{\nu}}} v \quad \text{iff} \quad \langle J_{\rho(\bar{\nu})}, A(\nu)\rangle \models \phi(u, v, \vec{x}, p(\bar{\nu})).$$

Define $\bar{B} \subseteq J_{\rho(\bar{\nu})}$ by

$$u \in \bar{B} \quad \text{iff} \quad \langle J_{\rho(\bar{\nu})}, A(\bar{\nu})\rangle \models \phi(u, \vec{y}, p(\bar{\nu})).$$

It is easily seen that $\bar{B}$ is a branch through $T^{\bar{\nu}}$ and that $\sigma_{\bar{\nu}\nu}[\bar{B}] \subseteq B$.

Now, the branch $\bar{B}$ just defined is Σ_1-definable over $\langle J_{\rho(\bar{\nu})}, A(\bar{\nu})\rangle$, so there is (by condition (iii) above) a point $x \in T^{\bar{\tau}}_{\bar{\nu}}$ extending $\bar{B}$. We shall ensure that $\pi_{\bar{\tau}\tau}(x)$ extends B on T^{τ}_{ν}. (This is where we lose the full normality of our trees. We deal not with branches so much as with Σ_1-definitions of branches, and we must handle different definitions as if they gave rise to "different" branches, though this may not be the case, of course.)

As in section 3, the construction proceeds by cases.

Case 1. τ is a limit point in $\dashv$.
Set $T^{\tau} = \bigcup_{\bar{\tau} \dashv \tau} \pi_{\bar{\tau}\tau}[T^{\bar{\tau}}]$. Conditions (i) and (ii) are verified as in section 3. We check (iii) and (iv) simultaneously.

Suppose then that $\tau = s(\nu)$, where ν is a limit point of S_{α_τ}, and that T^{ν} is Σ_1-definable over $\langle J_{\rho(\nu)}, A(\nu)\rangle$. Let B be a branch through T^{ν} which is Σ_1-definable over $\langle J_{\rho(\nu)}, A(\nu)\rangle$. Then there are Σ_1-formulas ϕ, ψ of $\mathcal{L}(A)$ and elements $\vec{x}, \vec{y} \in \alpha_\nu$ such that for all $u, v \in J_{\rho(\nu)}$,

$$u <_{T^\nu} v \quad \text{iff} \quad \langle J_{\rho(\nu)}, A(\nu)\rangle \models \phi(u, v, \vec{x}, p(\nu)),$$

$$u \in B \quad \text{iff} \quad \langle J_{\rho(\nu)}, A(\nu)\rangle \models \psi(u, \vec{y}, p(\nu)).$$

Pick $\bar{\tau}_0 \dashv \tau$ with $\vec{x}, \vec{y} \in \alpha_{\bar{\tau}_0}$. Let $\bar{\tau}_0 = s(\bar{\nu}_0)$. Suppose $\bar{\tau}_0 \dashv \bar{\tau} \dashv \tau$, and let $\bar{\tau} = s(\bar{\nu})$. Then, clearly, for $u, v \in J_{\rho(\bar{\nu})}$,

$$u <_{T^{\bar{\nu}}} v \quad \text{iff} \quad \langle J_{\rho(\bar{\nu})}, A(\bar{\nu}_0)\rangle \models \phi(u, v, \vec{x}, p(\bar{\nu})).$$

Define $B_{\bar{\nu}} \subseteq J_{\rho(\bar{\nu})}$ by

$$u \in B_{\bar{\nu}} \quad \text{iff} \quad \langle J_{\rho(\bar{\nu})}, A(\bar{\nu})\rangle \models \psi(u, \vec{y}, p(\bar{\nu})).$$

Thus $B_{\bar{\nu}}$ is a branch through $T^{\bar{\nu}}$. Moreover, for $\bar{\tau}_0 \dashv \bar{\tau} \dashv \bar{\tau}' \dashv \tau$, $\pi_{\bar{\nu}\bar{\nu}'}[B_{\bar{\nu}}] = B_{\bar{\nu}'} \cap \text{ran}(\pi_{\bar{\nu}\bar{\nu}'})$. Also, $\pi_{\bar{\nu}\nu}[B_{\bar{\nu}}] = B \cap \text{ran}(\pi_{\bar{\nu}\nu})$, and $B = \bigcup_{\bar{\tau}_0 \dashv \bar{\tau} \dashv \tau} \pi_{\bar{\nu}\nu}[B_{\bar{\nu}}]$. By condition (iii) below τ, let $x_{\bar{\nu}}$ extend $B_{\bar{\nu}}$ on

$T^{\bar{\tau}}_{\bar{\nu}}$. By condition (iv) below τ, $\pi_{\bar{\bar{\tau}}\bar{\tau}'}(x_{\bar{\nu}}) = x_{\bar{\nu}'}$ for $\bar{\tau}_0 \dashv \bar{\tau} \dashv \bar{\tau}' \dashv \tau$. Let $x = \pi_{\bar{\tau}\tau}(x_{\bar{\nu}})$ for any $\bar{\tau}, \bar{\tau}_0 \dashv \bar{\tau} \dashv \tau$. Clearly x extends B on T^{τ}_{ν}. This proves (iii). And the method of proof establishes (iv).

<u>Case 2.</u> τ is minimal in $\dashv$.

There are three subcases to consider.

<u>Case 2.1</u> τ is initial in S_{α_τ}.

Set $T^{\tau} = \phi$. There are no checks to be made.

<u>Case 2.2</u> τ is a limit point of S_{α_τ}.

Set $T^{\tau} = \cup\{T^{\nu} \mid \nu \in S_{\alpha_\tau} \cap \tau\}$. There is nothing to check.

<u>Case 2.3</u> $\tau = s(\nu)$.

There are three subcases to consider.

<u>Case 2.3.1</u> ν is initial in S_{α_τ}.

Set $T^{\tau}_{\nu} = \tau - \nu$ and finish.

<u>Case 2.3.2</u> $\nu = s(\eta)$.

Use the ordinals of $\tau - \nu$ to provide infinitely many extensions on T^{τ}_{ν} of each member of T^{ν}_{η}. There is nothing to check.

<u>Case 2.3.3</u> ν is a limit point of S_{α_τ}.

For each $x \in T^{\nu}$, let b_x be a branch through T^{ν} containing x. We shall use the ordinals from $\tau - \nu$ to provide extensions of each of these branches b_x on T^{τ}_{ν}, and possibly some other (countably many) branches. This latter possibility only arises when T^{ν} is Σ_1-definable over $\langle J_{\rho(\nu)}, A(\nu)\rangle$. In this case we extend each branch through T^{ν} which is Σ_1-definable over $\langle J_{\rho(\nu)}, A(\nu)\rangle$, treating distinct Σ_1-definitions as though they produced distinct branches, and thereby associating extensions with definitions rather than actual branches. (Hence, some branches may have more than one extension.) There are no checks to be made.

Case 3. τ immediately succeeds $\bar{\tau}$ in $\dashv$.

As in Case 2 there are three subcases.

Case 3.1 τ is initial in S_{α_τ}.

Set $T^\tau = \phi$ and be done.

Case 3.2 τ is a limit point of S_{α_τ}.

Set $T^\tau = \bigcup\{T^\nu | \nu \in S_{\alpha_\tau} \cap \tau\}$. The only non-trivial check to be made concerns condition (ii). This is handled as in section 3.

Case 3.3 $\tau = s(\nu)$.

Then $\bar{\tau} = s(\bar{\nu})$, where $\pi_{\bar{\tau}\tau}(\bar{\nu}) = \nu$. There are three subcases to consider.

Case 3.3.1 ν is initial in S_{α_τ}.

Let $T^\tau_\nu = \tau - \nu$. There are no non-trivial checks to be made.

Case 3.3.2 $\nu = s(\eta)$.

Then $\bar{\nu} = s(\bar{\eta})$, where $\pi_{\bar{\tau}\tau}(\bar{\eta}) = \eta$. For each pair $x,y \in T^{\bar{\tau}}$ such that $x \in T^{\bar{\tau}}_{\bar{\eta}}$, $y \in T^{\bar{\tau}}_{\bar{\nu}}$, and $x <_T y$, let $\pi_{\bar{\tau}\tau}(y)$ extend $\pi_{\bar{\tau}\tau}(x)$ on T^τ_ν. Use the ordinals in $\tau - (\nu \cup \pi_{\bar{\tau}\tau}[T^{\bar{\tau}}_{\bar{\nu}}])$ to ensure now that every member of T^ν_η has infinitely many extensions on T^τ_ν.

Case 3.3.3 ν is a limit point of S_{α_τ}.

There are two subcases to consider.

Case 3.3.3.1 $\pi_{\bar{\nu}\nu}[\bar{\nu}]$ is cofinal in ν.

For each $x \in T^{\bar{\tau}}_{\bar{\nu}}$, let $\pi_{\bar{\tau}\tau}(x)$ be an extension on T^τ_ν of the branch through T^ν determined by $\{\pi_{\bar{\tau}\tau}(y) | y <_T x\}$. Using the ordinals of $\tau - (\nu \cup \pi_{\bar{\tau}\tau}[T^{\bar{\tau}}_{\bar{\nu}}])$, we extend up to three further collections of branches through T^ν.

Firstly, if $\gamma \dashv \nu$ and $\gamma \neq \max(S_{\alpha_\gamma})$ and $\pi_{\gamma\nu}[\gamma]$ is cofinal in ν, then for each $x \in T_\gamma^{s(\gamma)}$ we ensure that the branch $\{\pi_{\gamma\nu}(y) | y <_T x\}^\frown$ through T^ν extends on T_ν^τ.

Secondly, if T^ν is Σ_1-definable over $\langle J_{\rho(\nu)}, A(\nu)\rangle$, we ensure that every branch through T^ν which is Σ_1-definable over $\langle J_{\rho(\nu)}, A(\nu)\rangle$ has an extension on T_ν^τ, regarding distinct Σ_1-definitions as if they defined distinct branches, as in Case 2.3.3 above.

Finally, we ensure that each point of T^ν lies on at least one branch through T^ν which is extended on T_ν^τ.

The only significant point to check is condition (iv) for the pair $\tau, \bar{\tau}$. But a few moments reflection suffice to show that there is no problem here, so we shall not give the simple details.

<u>Case 3.3.3.2</u> $\lambda = \sup \pi_{\bar{\nu}\nu}[\bar{\nu}] < \nu$.
Then $\lambda \in S_{\alpha_\tau}$, $\bar{\nu} \dashv \lambda$, and $\sigma_{\bar{\nu}\nu} \upharpoonright J_{\bar{\nu}} = \sigma_{\bar{\nu}\nu} \upharpoonright J_{\bar{\nu}}$. And, as in section 3 in this case we can easily show that for each $x \in T_{\bar{\nu}}^{\bar{\tau}}$ there is a point $y'(x)$ in T_λ^ν which extends the branch $\{\pi_{\bar{\tau}\tau}(z) | z <_T x\}^\frown$ through T^λ.

Suppose first that there is a Σ_1-formula ϕ of $\mathcal{L}(A)$ and ordinals $\vec{x} \in \alpha_{\bar{\nu}}$ such that for $u, v \in J_{\rho(\nu)}$,

$$u <_{T^\nu} v \quad \text{iff} \quad \langle J_{\rho(\nu)}, A(\nu)\rangle \models \phi(u, v, \vec{x}, p(\nu)).$$

Thus, for $u, v \in J_{\rho(\bar{\nu})}$, we have

$$u <_{T^{\bar{\nu}}} v \quad \text{iff} \quad \langle J_{\rho(\bar{\nu})}, A(\bar{\nu})\rangle \models \Phi(u, v, \vec{x}, p(\bar{\nu})).$$

Suppose further that ψ is a Σ_1-formula of $\mathcal{L}(A)$ and $\vec{y} \in \alpha_{\bar{\nu}}$ are such that B is a branch through T^ν, where $B \subseteq J_{\rho(\nu)}$ is defined by

$$u \in B \quad \text{iff} \quad \langle J_{\rho(\nu)}, A(\nu)\rangle \models \psi(u, \vec{y}, p(\nu)).$$

Then $\bar{B}$ is a branch through $T^{\bar{\nu}}$, where $\bar{B} \subseteq J_{\rho(\bar{\nu})}$ is defined by

$$u \in \bar{B} \quad \text{iff} \quad \langle J_{\rho(\bar{\nu})}, A(\bar{\nu})\rangle \models \psi(u, \vec{y}, p(\bar{\nu})).$$

We know that $\bar{B}$ has an extension, let us call it w, on $T^{\bar{\tau}}_{\bar{\nu}}$, associated with the above definition. Let $\pi_{\bar{\tau}\tau}(w)$ extend B on T^{τ}_{ν} (this extension being the one now to be associated with the above definition of B).

We complete the definition of T^{τ}_{ν} as follows. For each point $x \in T^{\bar{\tau}}_{\bar{\nu}}$ not considered above, choose a branch through T^{ν} containing y'(x) and let $\pi_{\bar{\tau}\tau}(x)$ extend that branch on T^{τ}_{ν}. Now, using the remaining ordinals in $\tau - \nu$, ensure that each point of T^{ν} has an extension on T^{τ}_{ν} and that, in case T^{ν} is Σ_1-definable over $\langle J_{\rho(\nu)}, A(\nu)\rangle$, every branch through T^{ν} which is Σ_1-definable over $\langle J_{\rho(\nu)}, A(\nu)\rangle$ extends onto T^{τ}_{ν}, where, as usual, we regard distinct definitions as if they defined distinct branches.

There are no non-trivial checks to be made.

The construction is complete. Clearly, $T = \bigcup_{\tau \in S_{\omega_1}} T^{\tau}$ is an almost normal tree of height ω_2 and width $\aleph_2$. We show that T has $\aleph_3$ many ω_2-branches. Suppose not, and let $B = \langle b_\xi \mid \xi < \omega_2\rangle$ enumerate all the ω_2-branches of T.

(A special case of the following argument shows that T has at least one ω_2-branch. Since the sequence $\langle b_\xi \mid \xi < \omega_2\rangle$ does not need to be one-one, this will exhaust all possibilities.)

Define submodels $N_\lambda \prec J_{\omega_3}$ as follows, for $\lambda < \omega_2$.

Let $N_o \prec J_{\omega_3}$ be smallest such that $\omega_1 \cup \{T,B\} \subseteq N_o$. Clearly $N_o \cap \omega_2$ is transitive. Set $\nu_o = N_o \cap \omega_2$.

Let $N_{\lambda+1} \prec J_{\omega_3}$ be smallest such that $N_\lambda \cup \{N_\lambda\} \subseteq N_{\lambda+1}$. Then $N_{\lambda+1} \cap \omega_2$ is transitive. Set $\nu_{\lambda+1} = N_{\lambda+1} \cap \omega_2$.

If $\lim(\lambda)$, let $N_\lambda = \bigcup_{\eta<\lambda} N_\eta$. Thus $N_\lambda \prec J_{\omega_3}$ and $N_\lambda \cap \omega_2 = \sup_{\eta<\lambda} \nu_\eta$.

Set $\nu_\lambda = \sup_{\eta<\lambda} \nu_\eta$.

Let $\rho_\lambda: N_\lambda \simeq J_{\kappa(\lambda)}$. Then $\rho_\lambda \upharpoonright \nu_\lambda = \mathrm{id}\upharpoonright \nu_\lambda$ and $\rho_\lambda(\omega_2) = \nu_\lambda$. It is easily seen that ν_λ is a limit point of S_{ω_1}. Moreover, $\rho_\lambda(T) = T^{\nu_\lambda}$ and $\rho_\lambda(B) = \langle b_\xi \cap T^{\nu_\lambda} | \xi < \nu_\lambda \rangle$.

We shall now attempt to define an ω_2-branch of T, distinct from each of b_ξ, $\xi < \omega_2$. If we succeed, we shall have arrived at our desired contradiction.

Let x_o be the least ordinal in T_{ν_o} not in b_o. If $x_\lambda \in T_{\nu_\lambda}$ is defined, let $x_{\lambda+1}$ be the least ordinal in $T_{\nu_{\lambda+1}}$ such that $x_\lambda <_T x_{\lambda+1}$ and $x_{\lambda+1} \notin b_\lambda$. If $\lim(\lambda)$ and $\langle x_\eta | \eta < \lambda \rangle$ is defined, let x_λ be the least ordinal in T_{ν_λ} which extends each of x_η, $\eta<\lambda$, if such an x_λ exists; otherwise the definition breaks down. Providing x_λ is defined for each $\lambda < \omega_2$, the sequence $\langle x_\lambda | \lambda < \omega_2 \rangle$ clearly defines an ω_2-branch of T distinct from each of b_ξ, $\xi < \omega_2$. So, we must show that if $\lim(\lambda)$ and $\langle x_\eta | \eta < \lambda \rangle$ is defined, then x_λ is also defined.

<u>Lemma 4.2</u>

For each $\lambda < \omega_2$, T^{ν_λ} is Σ_1-definable over $\langle J_{\rho(\nu_\lambda)}, A(\nu_\lambda) \rangle$.

Proof: Let $\lambda < \omega_2$, and set $\nu = \nu_\lambda$. Now, $\rho_\lambda^{-1}: J_{\kappa(\lambda)} \prec J_{\omega_3}$ and $\rho_\lambda^{-1}(\nu) = \omega_2$. Hence ν is regular over $J_{\kappa(\lambda)}$. But ν is not regular over $J_{\beta(\lambda)}$ (by definition of $\beta(\nu)$). Hence $\kappa(\lambda) < \beta(\nu)$. But $T^\nu =$ $= \rho_\lambda(T) \in J_{\kappa(\lambda)}$. Thus $T^\nu \in J_{\beta(\nu)}$. In particular, T^ν is $\Sigma_{n(\nu)}$-definable (with parameters) over $J_{\beta(\nu)}$. But $T^\nu \subseteq J_{\rho(\nu)}$. Thus, by the properties of the codes $A(\nu)$, T^ν is Σ_1-definable over $\langle J_{\rho(\nu)}, A(\nu) \rangle$, as required. □

By the above lemma and our construction (in particular, condition (iii)), it suffices now to show that for $\lim(\lambda)$, $\langle x_\eta | \eta < \lambda \rangle$ is Σ_1-definable over $\langle J_{\rho(\nu_\lambda)}, A(\nu_\lambda) \rangle$. And by the same considerations as above, this will follow at once if we can show that

$\langle x_\eta | \eta < \lambda \rangle \in J_{\beta(\nu_\lambda)}$. It clearly suffices to show that (setting $\nu = \nu_\lambda$ from now on) T^ν, $\langle b_\xi \cap T^\nu | \xi < \lambda \rangle$, $\langle \nu_\eta | \eta < \lambda \rangle \in J_{\beta(\nu)}$.

We know already that $T^\nu \in J_{\beta(\nu)}$. Also, $\langle b_\xi \cap T^\nu | \xi < \lambda \rangle = \langle b_\xi \cap T^\nu | \xi < \nu \rangle \restriction \lambda$ and $\langle b_\xi \cap T^\nu | \xi < \nu \rangle = \rho_\lambda(B) \in J_{\kappa(\lambda)} \subseteq J_{\beta(\nu)}$. So it remains to show that $\langle \nu_\eta | \eta < \lambda \rangle \in J_{\beta(\nu)}$.

Define N_η', $\eta < \lambda$, from $J_{\kappa(\lambda)}, \omega_1, T^\nu, \langle b_\xi \cap T^\nu | \xi < \nu \rangle$ just as N_η, $\eta < \omega_2$, were defined from $J_{\omega_3}, \omega_1, T, B$. Set $\nu_\eta' = N_\eta' \cap \nu$ for each $\eta < \lambda$. Then, $\langle N_\eta' | \eta < \lambda \rangle \in J_{\beta(\nu)}$, so $\langle \nu_\eta' | \eta < \lambda \rangle \in J_{\beta(\nu)}$.

(Recall that $\kappa(\lambda) < \beta(\nu)$.) But we may replace J_{ω_3} by N_λ in the original definition of N_η for $\eta < \lambda$. So, as $\rho_\lambda : N_\lambda \cong J_{\kappa(\lambda)}$, a simple induction shows that $(\rho_\lambda \restriction N_\eta) : N_\eta \cong N_\eta'$, and hence that $\nu_\eta = \nu_\eta'$, for all $\eta < \lambda$. Hence $\langle \nu_\eta | \nu < \lambda \rangle \in J_{\beta(\nu)}$, and we are done. □

References

[De1]. K.J. Devlin, Aspects of Constructibility. Springer Lecture Notes 354 (1973).

[De2]. -----------, Constructibility. Springer, to appear.

[Je]. T.J.Jech. Trees. Journal of Symbolic Logic 36, (1971), 1-14.

[La]. R. Laver & S. Shelah. The $\aleph_2$-Souslin Hypothesis.

[Mi]. W.J. Mitchell, Aronszajn Trees and the Independence of the Transfer Property. Annals of Math. Logic 5, (1972), 21-46.

[Si]. J.H. Silver, The Independence of Kurepa's Conjecture and Two-Cardinal Conjectures in Model Theory. AMS Proc. Symp. in Pure Maths. XIII, Part I, 383-390.

COARSE MORASSES IN L

Hans-Dieter Donder
Mathematisches Institut
Universität Bonn, BRD

Introduction Higher-gap morasses have been introduced by Jensen several years ago (see [5]). He used them to prove strong model-theoretic transfer theorems in L. This work is still unpublished but a definition of higher-gap morasses can be found in [6]. The proof that morasses exist in L essentially uses the fine structure of L. But Jensen noticed that weaker structures can be obtained rather easily just using coarse definability and condensation arguments in L. In his notes, Jensen gave a thorough axiomatic treatment of these structures. Our approach in § 2 is slightly different. We just define the "natural" global coarse morass in L and use its properties. One reason for this approach is the fact that Jensen has only worked out the axiomatic treatment for small gaps. This restriction is not necessary for the simple questions we deal with. In addition, we use some properties of the natural morass which do not seem to follow from the axioms and some which are actually false for fine morasses.

The main aim of this paper is to try to convince the reader that the coarse morasses in L are the natural tool to answer combinatorial questions in L which can be proved by coarse methods but not some direct application of a $\diamondsuit$-principle. In this direction, we show in § 2 that using the morass one can get Kurepa trees with additional properties. We also give some applications. In § 1 we only deal with gap-1 coarse morasses and among other things prove a simple combinatorial result which seems to be new. This result could also be proved using an appropriate $\diamondsuit$-principle but we think that the morass proof can be more easily visualized.

§ 1. Coarse gap-1 morasses

Gap-1 morasses have been described by Devlin in his book. Assuming for the moment that the reader has a copy of that book available we can define coarse gap-1 morasses as a structure which satisfies the morass axioms (M0) - (M5) (see [1], p.149). We first show that assuming $V=L$, of course, such a structure can be obtained very easily. Thereafter, we give some applications.

Assume $V=L$. Let $\tilde{S}$ be the class of all ordinals γ such that L_γ is a model of ZF^-. Let S be the class of limit points of $\tilde{S}$. In addition, set

$$S^+ = \{\nu \in S \mid L_\nu \models \text{"there is a largest uncountable cardinal"}\}$$

For $\nu \in S^+$ set

$$\alpha_\nu = \text{the largest } L_\nu\text{-cardinal}$$

and let

$$S_\alpha = \{\nu \in S^+ \mid \alpha_\nu = \alpha\}$$

Now let $\nu \in S^+$ such that ν is not a cardinal. We define

ν^* = the least $\tilde{\nu} \in \tilde{S}$ such that $\nu \leq \tilde{\nu}$
and for some $p \in L_{\tilde{\nu}}$ every $x \in L_{\tilde{\nu}}$
is $L_{\tilde{\nu}}$-definable from parameters
in $\alpha_\nu \cup \{p\}$

p_ν = the $<_L$-least such p

In addition set $q_\nu = \langle p_\nu, \nu \rangle$ if $\nu < \nu^*$ and $q_\nu = p_\nu$ otherwise.

To see that such a $\tilde{\nu}$ exists, let $\tau > \nu$ be a regular cardinal. Since ν is not a cardinal there is some $f \in L_\tau$ which maps α_ν onto ν. Let M be the elementary submodel of L_τ generated by $\alpha_\nu \cup \{f\}$. Let $M \cong L_{\tilde{\nu}}$. Then $\tilde{\nu}$ is as required.

The same kind of argument yields

__Remark 1:__ Let $\nu \in S_\alpha \cap \tau$. Then $\nu^* < \tau$.

The following basic lemma is immediately verified by standard definability arguments.

Lemma 1: Let $h : L_{\rho} \longrightarrow_{\Sigma_\omega} L_{\nu^*}$ such that $q_\nu \in \mathrm{rng}(h)$. Let $h(\bar\nu)=\nu$, if $\nu<\nu^*$, or $\bar\nu=\rho$ otherwise. Then $\bar\nu\in S^+$, $\rho=\bar\nu^*$ and $h(q_{\bar\nu})=q_\nu$.

We now define a relation $\prec$ on S^+ as follows:

Let $\bar\nu,\nu\in S^+$, $\bar\nu\neq\nu$

$\bar\nu\prec\nu$ iff there exists $f : L_{\bar\nu^*} \longrightarrow_{\Sigma_\omega} L_{\nu^*}$

such that $f\upharpoonright\alpha_{\bar\nu} = \mathrm{id}\upharpoonright\alpha_{\bar\nu}$ and $q_\nu\in \mathrm{rng}(f)$

Let $\bar\nu\prec\nu$ and f be as above. Then f is uniquely determined by $\bar\nu,\nu$. So we may set $\Pi_{\bar\nu\nu} = f\upharpoonright\bar\nu$. Sometimes we also use $\Pi_{\bar\nu\nu}$ to denote $f\upharpoonright L_{\bar\nu}$. We also set $\Pi_{\bar\nu\nu}(\bar\nu) = \nu$.

Obviously, $\prec$ is a tree, $\langle\Pi_{\nu\tau}\,|\,\nu\prec\tau\rangle$ is a commutative system and we have

$$\bar\nu\prec\nu \rightarrow \alpha_{\bar\nu}<\alpha_\nu$$

Now let us fix a regular cardinal κ and restrict our attention to points in $S^+\cap\kappa^+$.

Set $\mathcal{S} = \{\langle\alpha,\nu\rangle\,|\,\nu\in S\cap\kappa^+,\ \alpha=\alpha_\nu,\ L_\nu\models \text{"}\alpha \text{ is regular"}\}$

$S^o = \{\alpha\,|\,\exists\nu\ \langle\alpha,\nu\rangle\in\mathcal{S}\}$

$S^+ = \{\nu\,|\,\exists\alpha\ \langle\alpha,\nu\rangle\in\mathcal{S}\}$

Note that

$\langle\alpha,\nu\rangle,\langle\alpha',\nu'\rangle\in\mathcal{S},\ \alpha<\alpha' \rightarrow \nu<\alpha'$

The following properties are immediately verified.

(MO) (a) S_α is closed for $\alpha<\kappa$;

$S_\alpha\subset\alpha^++1$; S_α is bounded in α^+, if α is not a cardinal

(b) $\kappa = \max S^o = \sup(S^o\cap\kappa)$;

S_κ is club in κ^+

(M1) If $\nu \prec \tau$, then $\Pi_{\nu\tau} \upharpoonright \alpha_\nu = id \upharpoonright \alpha_\nu$,

$\Pi_{\nu\tau}(\alpha_\nu) = \alpha_\tau$ and

$\Pi_{\nu\tau} : \langle \nu+1, <, S_{\alpha_\nu} \rangle \longrightarrow_{\Sigma_0} \langle \tau+1, <, S_{\alpha_\tau} \rangle$

(M2) $\bar{\tau} \prec \tau$ and $\bar{\nu} \in S_{\alpha_{\bar{\tau}}} \cap \bar{\tau}$, $\nu = \Pi_{\bar{\tau}\tau}(\bar{\nu})$

$\curvearrowright \bar{\nu} \prec \nu$ and $\Pi_{\bar{\nu}\nu} \upharpoonright \bar{\nu} = \Pi_{\bar{\tau}\tau} \upharpoonright \bar{\nu}$

(M3) $\{\alpha_\nu \mid \nu \prec \tau\}$ is closed in α_τ

(M4) τ not maximal in S_{α_τ} .

$\curvearrowright \{\alpha_\nu \mid \nu \prec \tau\}$ unbounded in α_τ

[To prove this, take $\rho \in S_{\alpha_\tau}$ such that $\tau < \rho$. Then $\tau^* < \rho$ by Remark 1. Choose, $\tau^* < \gamma < \rho$ such that $L_\gamma \models ZF^-$. Given any $\beta < \alpha_\tau$ we find working in L_γ where α_τ is regular some $X \prec L_{\tau^*}$ such that $\beta \in X \cap \alpha_\tau \in \alpha_\tau$ and $q_\tau \in X$.
But then collapsing X and applying Lemma 1 we get $\nu \prec \tau$ such that $\alpha_\nu = X \cap \alpha_\tau$]

(M5) $\{\alpha_\nu \mid \nu \prec \tau\}$ unbounded in α_τ

$$\tau = \bigcup_{\nu \prec \tau} \Pi_{\nu\tau}'' \nu$$

We now turn to some applications. For notational simplicity we restrict ourselves to the case $\kappa = \omega_1$. The generalizations to arbitrary successor cardinals will be obvious. Inaccessible cardinals κ will be discussed at the end of this chapter.

We shall actually work with the natural gap-1 morass defined above i.e. we use some properties which are not given by the axioms (M0) – (M5). In particular, we essentially use the fact that our morass is universal.

We first introduce some notations. For $\nu \in S^+$ set $A_\nu = \{\alpha_{\bar{\nu}} \mid \bar{\nu} \prec \nu\}$ and $B_\nu = \{\bar{\nu} \mid \bar{\nu} \prec \nu\}$. Note that $B_\nu, A_\nu \in L_\delta$ when ν is not a cardinal in L_δ and δ is closed enough. So especially we get

Lemma 2: $\nu\in S^+,\ \nu<\tau,\ \tau\in S^0\cup S^1 \rightarrow A_\nu, B_\nu \in L_\tau$

The following strengthening of $\diamondsuit^+$ was introduced by Devlin in [3].

$\diamondsuit^\#$: there is a sequence $\langle N_\alpha | \alpha<\omega_1\rangle$ s.t.

(i) N_α is a countable, transitive p.r. closed set containing α

(ii) if $X\subseteq\omega_1$ there is a club $C\subseteq\omega_1$ s.t. $\alpha\in C \rightarrow X\cap\alpha,\ C\cap\alpha\in N_\alpha$

(iii) $\langle N_\alpha | \alpha<\omega_1\rangle$ is Π^1_n-reflecting for $n<\omega$ which means:

whenever a Π^1_n-sentence Φ is true in a structure $\langle\omega_1, \in, (A_i)_{i<\omega}\rangle$,

then there is an $\alpha<\omega_1$ s.t.

$N_\alpha \models$ "Φ is true in $\langle\alpha, \in, (A_i\restriction\alpha)_{i<\omega}\rangle$"

[Actually, Devlin only requires Π^1_2-reflection]

We now show that a natural $\diamondsuit^\#$- sequence is contained in the morass.

Let $\alpha<\omega_1$. Set $\delta(\alpha) = \max(S_\alpha\cup\{\alpha\})$ if $\delta(\alpha)$ is not a limit point in $\prec$, and set $\delta(\alpha) = \min(S^0-(\alpha+1))$ otherwise.

By Lemma 2 we have

(*) $\nu\in S_\alpha \rightarrow A_\nu\in N_\alpha$

Lemma 3: $\langle N_\alpha | \alpha<\omega_1\rangle$ satisfies $\diamondsuit^\#$

Proof: We have to verify (i) - (iii) in the definition of $\diamondsuit^\#$.

(i) is obvious. (ii) Let $X\subseteq\omega_1$. Choose $\nu\in S_{\omega_1}$ s.t. $X\in L_\nu$. There is some $\bar\nu<\nu$ s.t. $X\in \operatorname{rng}\Pi_{\bar\nu\nu}$. Then (*) shows that $A_\nu\cap\alpha_{\bar\nu}$ is as required.

(iii) Let $\mathfrak{A} = \langle\omega_1, \in, (A_i)_{i<\omega}\rangle$ be given and let Φ be a Π^1_n-sentence which is true in $\mathfrak{A}$. Since S_{ω_1} is club in ω_2 there is some $\nu\in S_{\omega_1}$ s.t. $L_\nu \models$ "Φ is true in $\mathfrak{A}$". Choose $\bar\nu\prec\nu$ s.t. $\mathfrak{A}\in \operatorname{rng}\Pi_{\bar\nu\nu}$ and $\bar\nu$ is a successor in $\prec$. Since $\Pi_{\bar\nu\nu}: L_{\bar\nu} \longrightarrow_{\Sigma_\omega} L_\nu$ we clearly have

$L_{\bar\nu} \models$ "Φ is true in $\mathfrak{A}\restriction\alpha_{\bar\nu}$" .

But $\bar\nu$ is a successor in $\prec$. Hence $\bar\nu = \max S_{\alpha_{\bar\nu}}$.

So $N_{\alpha_{\bar\nu}} = L_{\bar\nu}$ qued

Devlin showed in [3] that $\diamondsuit^{\#}$ implies the existence of a Kurepa tree without Aronszajn subtrees. Nevertheless we shall show now that the morass gives us a "natural" tree of that kind. We shall use this fact in § 2, where we shall show that the corresponding tree for $\kappa = \lambda^{+} > \omega_1$ has additional nice properties. The morass tree itself is not good enough, for there are points $\nu \in S' \cap \omega_1$ which have ω_1 many immediate successors. But it is easy to see that $\{B_\nu \mid \nu \in S_{\omega_1}\}$ is a Kurepa family. We shall take the tree associated with that family.

For $\nu \in S'$ let $g_\nu : \alpha_\nu \longrightarrow 2$ be the characteristic function of B_ν (i.e. $g_\nu(\delta) = 0$ iff $\delta \in B_\nu$). Set

$T = \{g_\nu \restriction \delta \mid \nu \in S_{\omega_1}, \delta < \omega_1\}$ and let $\underset{\sim}{T} = \langle T, \subseteq \rangle$

<u>Lemma 4</u>: $\underset{\sim}{T}$ is a Kurepa tree which contains no Aronszajn subtree.

<u>Proof</u>: For the first part it suffices to show that $F = \{B_\nu \mid \nu \in S_{\omega_1}\}$ is a Kurepa family. It is easy to see that $B_\nu \neq B_{\bar\nu}$ for $\nu, \bar\nu \in S_{\omega_1}, \nu \neq \bar\nu$. But for $\alpha < \omega_1$ we have

$$F \restriction \alpha \subseteq \{B_\tau \mid \tau \in S_\alpha\} \cup \{B_\tau \cup \{\tau\} \mid \tau \in S' \cap \alpha\} \cup \{\emptyset\}$$

Hence $F \restriction \alpha$ is countable.

Now let $\bar T \subseteq T$ s.t. $\underset{\sim}{\bar T} = \langle \bar T, \subseteq \rangle$ is an ω_1-tree. We have to show that $\bar T$ has an uncountable branch. Choose $\nu \in S_{\omega_1}$ s.t. $\bar T \in L_\nu$ and let $\bar\nu \prec \nu$ s.t. $\bar\nu$ is a successor in $\prec$ and $\Pi_{\bar\nu\nu}(\bar T') = \bar T$ for some $\bar T'$. Let $\alpha = \alpha_{\bar\nu}$. Since $\bar T$ is an ω_1-tree and $\Pi_{\bar\nu\nu}$ is elementary we clearly have $\bar T' = \bigcup_{\beta < \alpha} \bar T_\beta$. So there is some $g \in T_\alpha$ s.t.

$\{\delta < \alpha \mid g \restriction \delta \in \bar T\}$ is unbounded in α. But our description of T_α above shows that $T_\alpha \subseteq L_{\bar\nu}$ (since $\bar\nu$ is a successor in $\prec$, hence $\bar\nu = \max S_\alpha$). So $g \in L_{\bar\nu}$. But then $\Pi_{\bar\nu\nu}(g)$ describes an uncountable branch of $\bar T$.

qued.

As we said before, Lemma 3 and 4 have obvious generalizations to successor cardinals. For inaccessible κ the situation is different.

The reason is that we have $|S_\alpha|=\alpha^+$ if α is a cardinal. But we shall show that for many inaccessible κ we only need small segments of the S_α $(\alpha<\kappa)$ to approximate S_κ. This argument is due to Jensen.

We first recall a familiar definition.

Definition: $\kappa>\omega$ is ineffable iff for every $\langle A_\alpha|\alpha<\kappa\rangle$ s.t. $A_\alpha\subseteq\alpha$ there is some $A\subseteq\kappa$ s.t. $\{\alpha<\kappa|A_\alpha=A\cap\alpha\}$ is stationary.

Now let κ be regular but not ineffable. Let $\vec{A}=\langle A_\alpha|\alpha<\kappa\rangle$ be the $<_L$-least counterexample to ineffability. For $\alpha\leq\kappa$ set

$\tilde{S}_\alpha=\{\nu\in S_\alpha|\vec{A}\in L_\nu,\ L_\nu\models$ "$\vec{A}\restriction\alpha$ shows that α is not ineffable"$\}$

and $\bar{S}_\kappa=\tilde{S}_\kappa$, $\bar{S}_\alpha=\{\nu\in\tilde{S}_\alpha|A_\alpha\notin L_\nu\}$ $(\alpha<\kappa)$.

Clearly, $\bar{S}_\alpha$ is closed for $\alpha<\kappa$ and $\bar{S}_\kappa$ is club in κ^+ and $|\bar{S}_\alpha|\leq\alpha$ for $\alpha<\kappa$.

For $\bar{\nu},\nu\in\bigcup\limits_{\alpha\leq\kappa}\bar{S}_\alpha$ set

$$\bar{\nu}\prec_\kappa\nu \text{ iff } \bar{\nu}<\nu \text{ and } \vec{A}\restriction\alpha_\nu\in \operatorname{rng}\Pi_{\bar{\nu}\nu}$$

Lemma 5: Let $\bar{\nu}<\nu$, $\nu\in\bar{S}_\alpha$ s.t. $\vec{A}\restriction\alpha\in\operatorname{rng}\Pi_{\bar{\nu}\nu}$, and set $\bar{\alpha}=\alpha_\nu$. Then $\bar{\nu}\in\bar{S}_{\bar{\alpha}}$, hence $\bar{\nu}\prec_\kappa\nu$.

Proof: We clearly have $\Pi_{\bar{\nu}\nu}(\vec{A}\restriction\bar{\alpha})=\vec{A}\restriction\alpha$. So we only have to show that $A_{\bar{\alpha}}\notin L_{\bar{\nu}}$. So assume $A_{\bar{\alpha}}\in L_{\bar{\nu}}$. We shall derive a contradiction. Since $\bar{\nu}\in\tilde{S}_{\bar{\alpha}}$, there is a club $\bar{C}\subseteq\bar{\alpha}$, $\bar{C}\in L_\nu$ s.t. $A_\beta\neq A_{\bar{\alpha}}\cap\beta$ for all $\beta\in\bar{C}$. Now set $C=\Pi_{\bar{\nu}\nu}(\bar{C})$, $A=\Pi_{\bar{\nu}\nu}(A_{\bar{\alpha}})$. Then $\bar{\alpha}\in C$ and $A\cap\bar{\alpha}=A_{\bar{\alpha}}$. But the elementary embedding $\Pi_{\bar{\nu}\nu}$ also gives us that $A\cap\beta\neq A_\beta$ for all $\beta\in C$. Contradiction.

qued.

Lemma 5 shows that the previous arguments go through for $\prec_\kappa$. So we get

Lemma 6: (V=L). Let κ be regular but not ineffable. Then $\diamondsuit^{\#}_\kappa$ holds i.e. there is a sequence $\langle N_\alpha|\alpha<\kappa\rangle$ s.t.

(i) N_α is a transitive p.r. closed set containing α and $|N_\alpha| \leq \alpha$

(ii) if $X \subseteq \kappa$ there is a club $C \subseteq \kappa$ s.t. $\alpha \in C \rightarrow X \cap \alpha, C \cap \alpha \in N_\alpha$

(iii) $\langle N_\alpha | \alpha < \kappa \rangle$ is Π^1_n-reflecting for all $n < \omega$

It is well known that $\diamondsuit^*_\kappa$ is already false for ineffable κ. So the lemma above gives various characterizations of ineffable cardinals in L. We now give another one.

<u>Proposition 7</u>: (V=L) Let κ be regular but not ineffable. Then there are partitions $f_\nu : \kappa \longrightarrow 2$ $(\nu < \kappa^+)$ s.t.

(*) if $A \subseteq \kappa^+$, $|A| = \kappa$, there is a club $C \subseteq \kappa$ s.t. $\forall \alpha \in C \; \exists \nu, \mu \in A (f_\nu(\alpha) = 0$ and $f_\mu(\alpha) = 1)$

<u>Proof</u>: Let $\bar{S}_\alpha, \prec_\kappa$ be defined as above. For $\alpha < \kappa$ set

$$G_\alpha = \{A \subseteq \bar{S}_\alpha \mid |A| = |\alpha| \text{ and } A \in L_\nu \text{ for some } \nu \in \bar{S}_\alpha\}$$

Then $|G_\alpha| \leq \alpha$. So by an easy diagonalization we can define $h_\alpha : \bar{S}_\alpha \longrightarrow 2$ s.t.

$$\forall A \in G_\alpha \; \exists \nu, \mu \in A (h_\alpha(\nu) = 0 \text{ and } h_\alpha(\mu) = 1)$$

For $\nu \in \bar{S}_\kappa$ now define $f_\nu : \kappa \longrightarrow 2$ by

$$f_\nu(\alpha) = \begin{cases} h_\alpha(\bar{\nu}) & \text{if } \bar{\nu} \prec_\kappa \nu, \; \bar{\nu} \in \bar{S}_\alpha \\ 0 & \text{if there is no such } \bar{\nu} \end{cases}$$

Note that given α there is at most one such $\bar{\nu}$. Let $A \subseteq \bar{S}_\kappa$ s.t. $|A| = \kappa$. Choose $\tau \in \bar{S}_\kappa$ s.t. $A \in L_\tau$. Let $\tau' \prec_\kappa \tau$ s.t. $A \in \text{rng}\, \Pi_{\tau'\tau}$ and set $C = \{\alpha_{\bar{\tau}} \mid \tau' \prec_\kappa \bar{\tau} \prec_\kappa \tau\}$. Then C is club in κ. It is immediate that C satisfies the requirements for A.

qued.

The proof shows that if $\kappa = \lambda^+$ we can strengthen (*) to:

if $A \subseteq \kappa^+$, $|A| \geq \lambda$, then ...

Actually, Jensen has shown a long time ago that for $\kappa = \lambda^+$ one can strengthen (*) to: if $A \subseteq \kappa^+$, $|A| = \lambda$, there is some $\delta < \kappa$ s.t. $\forall \alpha \in \kappa - \delta \; \exists \nu, \mu \in A (f_\nu(\alpha) = 0$ and $f_\nu(\alpha) = 1)$. But the proof uses a fine morass.

For inaccessible κ the situation is slightly different. Namely, Wolsdorf and Choodnovsky have shown (see [8]):

Let κ be weakly compact. Let $f_\nu : \kappa \longrightarrow 2$ $(\nu<\kappa^+)$. Then there is some $A\subseteq\kappa^+$, $|A|=\kappa$, and some unbounded $B\subseteq\kappa$ s.t. for some $i<2$

$$\forall\nu\in A \quad \forall\alpha\in B \quad f_\nu(\alpha)=i$$

For the sake of completeness we now show that Lemma 7 characterizes ineffable cardinals in L. The following result is proved in ZFC.

<u>Proposition 8</u>: Let κ be ineffable. Let $f_\nu : \kappa \longrightarrow 2$ $(\nu<\kappa^+)$. Then there is some $A\subseteq\kappa^+$, $|A|=\kappa$, and some stationary $E\subseteq\kappa$ s.t. for some $i<2$ $\quad \forall\nu\in E \quad \forall\alpha\in A \quad f_\nu(\alpha)=i$

<u>Proof</u>: We first introduce some notation. For $G\in\mathfrak{P}(\kappa)$ s.t. $|G|\leq\kappa$ let ΔG be the diagonal intersection of some enumeration $\langle G_\alpha|\alpha<\kappa\rangle$ of $G\cup\{\kappa\}$. It is easy to see that the statement "ΔG is stationary" does not depend on the choice of the enumeration.

Now set $A_\nu^i = \{\alpha<\kappa|f_\nu(\alpha)=i\}$. It suffices to find some $B\subseteq\kappa^+$, $|B|=\kappa$, and $h : B \longrightarrow 2$ s.t.

(1) $\Delta\{A_\nu^{h(\nu)}|\nu\in B\}$ is stationary

and (2) if $|D|<\kappa$ and $D\subseteq A_\nu^{h(\nu)}$ for some $\nu\in B$,

then $|\{\nu\in B|D\subseteq A_\nu^{h(\nu)}\}|=\kappa$

For then it is easy to construct $A\subseteq B$,

$|A|=\kappa$, s.t. $\Delta\{A_\nu^{h(\nu)}|\nu\in A\} = \cap\{A_\nu^{h(\nu)}|\nu\in A\}$.

Now set $W = \{D\subseteq\kappa|\ |D|<\kappa$ and $D\subseteq A_\nu^i$ for some $i,\nu\}$ and

$$I_D = \{\nu<\kappa^+|D\subseteq A_\nu^i \text{ for some } i\}.$$

Since $|W|\leq\kappa$, we may assume w.l.o.g. that $|I_D|=\kappa^+$ for all $D\in W$. Now given D let $g=g_D$ s.t. $D\subseteq A_\nu^{1-g(\nu)}$ for $\nu\in I_D$ and set $\mathcal{F}_D^\rho = \{A_\nu^{g(\nu)}|\nu\in I_D-\rho\}$. Now observe that if there is some $\rho<\kappa^+$ and $D\in W$ s.t. ΔG is stationary for every $G\subseteq\mathcal{F}_D^\rho$, $|G|\leq\kappa$, we immediately find h satisfying (1), (2). So we may assume that this is not the case. Hence we find for every $D\in W$ pairwise disjoint $G_D^\delta \subseteq \kappa^+ (\delta<\kappa)$ s.t. $|G_D^\delta| \leq \kappa$ and $\Delta\{A_\nu^{g(\nu)} \mid \nu\in G_D^\delta\}$ is not stationary (where $g=g_D$). Now set $B = \bigcup\{G_D^\delta \mid D\in W,\ \delta<\kappa\}$. Since κ is ineffable, there is some $h : B \longrightarrow 2$ s.t. $\Delta\{A_\nu^{h(\nu)} \mid \nu\in B\}$ is stationary (see [9]). But then $h\restriction G_D^\delta \neq g_D\restriction G_D^\delta$ for every D,δ. So h satisfies (2), too.

qued.

§ 2. The global coarse morass in L

Assume $V=L$ again and let $\kappa>\omega$ be regular but not ineffable. Let $\bar{S}_\alpha(\alpha\leq\kappa)$ and $\prec_\kappa$ be defined as in § 1. Set $\bar{B}_\nu = \{\bar{\nu} \mid \bar{\nu}\prec_\kappa \nu\}$ for $\nu\in\bar{S}_\kappa$ and let $F_\kappa = \{\bar{B}_\nu \mid \nu\in\bar{S}_\kappa\}$. For $\nu\in\bar{S}_\kappa$ let g_ν be the characteristic function of $\bar{B}_\nu$ and set $T^\kappa = \{g_\nu\restriction\alpha \mid \nu\in\bar{S}_\kappa, \alpha<\kappa\}$ and $\underset{\sim}{T}^\kappa = \langle T^\kappa, \subseteq\rangle$. Our main aim in this chapter is to investigate the trees $\underset{\sim}{T}^\kappa$ more closely. For this we introduce a definition. First let us remark that for us a λ-tree is a tree T of height λ s.t. $|T_\alpha|<\lambda$ for all $\alpha<\lambda$.

Definition 1: Let $\kappa>\omega$ be regular and let $F\subseteq\mathfrak{P}(\kappa)$. For $B\in F$ let $g_B : \kappa \longrightarrow 2$ be the characteristic function of B. Set $T = T_F = \{g_B\restriction\alpha \mid B\in F,\ \alpha<\kappa\}$ and $\underset{\sim}{T} = \langle T, \subseteq\rangle$.
$\underset{\sim}{T}$ is called a rich κ-tree iff the following conditions are satisfied.

(i) $|F| = \kappa^+$

(ii) for all $X\subseteq\kappa$, $\omega\leq|X|<\kappa$: $|F\restriction X| \leq |X|$
where $F\restriction X = \{B\cap X \mid B\in F\}$

(iii) let $\bar{T}$ be a subtree of $\underset{\sim}{T}$ s.t. $\bar{T}$ is a λ-tree where $\lambda\leq\kappa$ is regular. Then $\bar{T}$ has a branch of length λ.

Clearly, this is not a very elegant definition, but it is good enough for our purposes.

We shall prove:

Theorem 1: Assume $V=L$. Let $\kappa>\omega$ be regular but not ineffable. Then there is a rich κ-tree.

In fact, we shall show that for κ as above the tree $\underset{\sim}{T}^{\kappa}$ is rich. Of course, we have already shown this for $\kappa=\omega_1$ in § 1. But for $\kappa>\omega_1$ we now have to investigate small subtrees of $\underset{\sim}{T}^{\kappa}$. Since this will be done in a uniform way, we do not have to fix κ in the beginning. So we introduce now the general framework which might be called the "global coarse morass in L". We use the notations Jensen has introduced in the context of his higher-gap morasses (see [5]), but we do not give an axiomatic treatment.

Let S, S^+, S_α, ν^*, p_ν, q_ν be defined as in § 1. For $\nu \in S^+$-Card set $\mu(\nu) = \max\{\mu \in S \mid L_\mu \models$ "ν is a cardinal"$\}$. We note that $\mu(\nu)$ can also be found as follows.

Define $\langle \mu(\nu,i) \mid i<k_\nu \rangle$ by

$$\mu(\nu,0) = \nu \qquad \mu(\nu,\xi+1) = \begin{cases} \max S_{\mu(\nu,\xi)} & \text{if } S_{\mu(\nu,\xi)} \neq \emptyset \\ \text{undefined otherwise} \end{cases}$$

$$\mu(\nu,\lambda) = \sup_{\xi<\lambda} \mu(\nu,\xi) \quad \text{if } \lim(\lambda)$$

Then $\mu(\nu,k_\nu) = \mu(\nu)$.

We obviously have $\mu(\nu) \le \nu^*$, since the definition of ν^* guarantees that ν is not a cardinal in L_{ν^*+2} .

<u>Definition:</u> Let $\bar{\nu}, \nu \in S^+$-Card and $f : \mu(\bar{\nu}) \longrightarrow \mu(\nu)$.
Then $f : \bar{\nu} \Longrightarrow \nu$ iff f has an extension f^* s.t. $f^* : L_{\bar{\nu}^*} \longrightarrow_{\Sigma_\omega} L_{\nu^*}$ and $q_\nu \in \operatorname{rng} f^*$.

An easy argument shows that $f^*(q_{\bar{\nu}}) = q_\nu$. So f^* is uniquely determined by $f \restriction \alpha_{\bar{\nu}}$ and $\bar{\nu}, \nu$. So, by slight abuse of notation, given $f : \bar{\nu} \Longrightarrow \nu$ we do not distinguish f and f^*. We also set $f(\bar{\nu}^*) = \nu^*$. Note that $\Pi_{\bar{\nu}\nu} : \bar{\nu} \Longrightarrow \nu$, if $\bar{\nu} \prec \nu$. We clearly have

(E 1) Let $f : \bar{\nu} \Longrightarrow \nu$. Then $f(k_{\bar{\nu}}) = k_\nu$ and
$f(\mu(\bar{\nu},i)) = \mu(\nu,f(i))$ for $i \le k_{\bar{\nu}}$.

We now introduce another notation.

Let $f : \bar{\nu} \Longrightarrow \nu$ and $\bar{\tau} \in S^+$ s.t. $\mu(\bar{\tau}) \le \mu(\bar{\nu})$. Set $\tau = f(\bar{\tau})$. Define $f^{(\bar{\tau})} : \mu(\bar{\tau}) \longrightarrow \mu(\tau)$ by $f^{(\bar{\tau})} = f \upharpoonright \mu(\bar{\tau})$. We have

(E 2) $f^{(\bar{\tau})} : \bar{\tau} \Longrightarrow \tau$

<u>Proof</u>: Since $\mu(\bar{\tau}) \le \mu(\bar{\nu})$, it is easy to see that $\bar{\tau}^* \le \bar{\nu}^*$. Now if $\bar{\tau}^* < \bar{\nu}^*$ we get $f(\langle \bar{\tau}^*, q_{\bar{\tau}} \rangle) = \langle \tau^*, q_\tau \rangle$ since f is elementary. Then the claim is immediate. So assume $\bar{\tau}^* = \bar{\nu}^*$. We immediately get $\tau^* = \nu^*$. so we only have to show that $p_\tau \in \operatorname{rng}(f)$. Assume first that $\alpha_\nu \le \alpha_\tau$. Then p_τ is the $<_L$-least $p \in L_{\nu^*}$ s.t. p_ν is definable in L_{ν^*} with parameters from $\alpha_\tau \cup \{p\}$, and the corresponding statement is true for $p_{\bar{\tau}}$. Hence an easy argument shows that $f(p_{\bar{\tau}}) = p_\tau$. If $\alpha_\tau < \alpha_\nu$, the proof is similar.

qued.

(E 3) Let $f : \bar{\nu} \Longrightarrow \nu$ and $\bar{\tau} \in S_{\bar{\alpha}} \cap (\bar{\nu}+1)$, where $\bar{\alpha} = \alpha_{\bar{\nu}}$. Let $\bar{\eta} \prec \bar{\tau}$. Then $f(\bar{\eta}) \prec f(\bar{\tau})$.

<u>Proof</u>: By (E 2) we may assume w.l.o.g. that $\bar{\tau} = \bar{\nu}$, since otherwise we can consider $f^{(\bar{\tau})}$. Now set $\bar{\delta} = \alpha_{\bar{\eta}}$, $\delta = f(\bar{\delta})$ and $\eta = f(\bar{\eta})$. Set

X = the Skolem hull of $\delta \cup \{q_\nu\}$ in L_{ν^*}

$\bar{X}$ = the Skolem hull of $\bar{\delta} \cup \{q_{\bar{\nu}}\}$ in $L_{\bar{\nu}^*}$

By § 1, Lemma 1 we only have to show that $X \cap \alpha_\nu = \delta$ and otp $(X \cap \nu) = \eta$. But be know that $\bar{X} \cap \alpha_{\bar{\nu}} = \bar{\delta}$ and otp $(\bar{X} \cap \bar{\nu}) = \bar{\eta}$. So the conclusion follows by applying f, since L_{ν^*} has definable Skolem functions.

qued.

We now introduce some special maps.

<u>Definition</u>: Let $\nu \in S^+$-Card, $\alpha \le \nu$ and $x \in L_{\mu(\nu)}$. Let

X = the Skolem hull of $\alpha \cup \{x, q_\nu\}$ in L_{ν^*} .

Let $\bar{f} : L_\rho \xrightarrow{\sim} X$. By § 1, Lemma 1 we know that $L_\rho = L_{\bar{\nu}^*}$, where $\bar{f}(\bar{\nu}) = \nu$. Set $f = \bar{f} \upharpoonright \mu(\bar{\nu})$. Then $f : \bar{\nu} \Longrightarrow \nu$. We set $f_{(\alpha, x, \nu)} := f$.

Obviously, the characteristic property of $f_{(a,x,\nu)}$ is

(E 4) Let $f_{(a,x,\nu)} : \bar{\nu} \Longrightarrow \nu$ and $g : \tau \Longrightarrow \nu$ s.t. $a \cup \{x\} \subseteq \mathrm{rng}(g)$. Then there is some $h : \bar{\nu} \Longrightarrow \tau$ s.t. $f_{(a,x,\nu)} = g \cdot h$. In addition, we have $|\bar{\nu}| \leq |a| + \omega$.

We now come to the main property for which we need another definition. Let $\langle \eta(\gamma) \mid \gamma < \rho \rangle$ be a $\prec$-chain. We say that $\langle \eta(\gamma) \rangle$ <u>converges</u> iff there is some η s.t. $\eta(\gamma) \prec \eta$ for all $\gamma < \rho$.

(E 5) Let $f : \bar{\nu} \Longrightarrow \nu$. Let $\langle \eta(\gamma) \mid \gamma < \rho \rangle$ be a convergent $\prec$-chain where $\lim(\rho)$ and let $\eta(\gamma) = f(\bar{\eta}(\gamma))$ $(\gamma < \rho)$. Then $\langle \bar{\eta}(\gamma) \mid \gamma < \rho \rangle$ is a convergent $\prec$-chain.

<u>Proof</u>: By (E 2) (E 3) we know that $\langle \bar{\eta}(\gamma) \mid \gamma < \rho \rangle$ is a $\prec$-chain. It is easy to see that $\Pi_{\eta(\gamma),\eta(\delta)} = f(\Pi_{\bar{\eta}(\gamma),\bar{\eta}(\delta)})$ for $\gamma < \delta < \rho$.

Now let $\langle \langle U, E \rangle , \bar{\Pi}_{\gamma} \rangle$ be the direct limit of $\langle L_{\bar{\eta}(\gamma)^*} , \Pi_{\bar{\eta}(\gamma),\bar{\eta}(\delta)} \rangle$. Choose η s.t. $\eta(\gamma) \prec \eta$ for all $\gamma < \rho$. We can then define an elementary embedding $h : \langle U, E \rangle \longrightarrow_{\Sigma_\omega} L_{\eta^*}$ s.t. the following diagrams commute

$$\begin{array}{ccc} L_{\eta(\gamma)^*} & \xrightarrow{\ \Pi_{\eta(\gamma),\eta}\ } & L_{\eta^*} \\ \uparrow f & & \uparrow h \\ L_{\eta(\gamma)^*} & \xrightarrow{\ \bar{\bar{\Pi}}_{\gamma}\ } & \langle U, E \rangle \end{array}$$

But then $\langle U, E \rangle$ is well-founded, hence we may assume w.l.o.g. that $\langle U, E \rangle = \langle L_\beta, \in \rangle$ for some β. We clearly have $q_\eta \in \mathrm{rng}(h)$. But then $\beta = \bar{\eta}^*$, where $h(\bar{\eta}) = \eta$. Clearly, $\bar{\eta}(\gamma) \prec \bar{\eta}$ for all $\gamma < \rho$.

qued.

To illustrate our methods we consider the two-cardinal version of $\diamondsuit^+$. Set $\mathfrak{P}_\lambda(\kappa) = \{a \subseteq \kappa \mid |a| < \lambda\}$. For $a \in \mathfrak{P}_\lambda(\kappa)$ set $\Theta(a) = \cup a$. We recall the following definition (see [1])

$\diamondsuit^{+}_{\kappa,\lambda}$: There is a sequence $\langle N_a | a \in \mathfrak{P}_\lambda(\kappa)\rangle$ s.t. $|N_a| \leq |a|+\omega$, and whenever $X \subseteq \kappa$ there is an unbounded set $B \subseteq \kappa$ s.t. for any $a \in \mathfrak{P}_\lambda(\kappa)$, if $\Theta = \Theta(a)$ is a limit point of $B \cap a$, then $X \cap \Theta$, $B \cap \Theta \in N_a$

Now let κ be regular but not ineffable. Let $\vec{A} = \langle A_\alpha | \alpha < \kappa\rangle$ be the sequence we used to define $\prec_\kappa$. Let $a \in \mathfrak{P}_\kappa(\kappa)$ and let $\Theta = \Theta(a)$ and set $x = x(a) = \langle a, \Theta, \vec{A}\rangle$. We first set $f_a = f_{(|a|,x,\nu)}$ where $\nu = \min \bar{S}_\kappa$. Note that $a \subseteq \mathrm{rng}(f_a)$ and $\mathrm{rng}(f_a) \leq |a|+\omega$.
We set $N_a = \mathrm{rng}(f_a)$.

Lemma 2: $\langle N_a | a \in \mathfrak{P}_\kappa(\kappa)\rangle$ satisfies $\diamondsuit^{+}_{\kappa\kappa}$

Proof: Assume not. Let $X \subseteq \kappa$ be the $<_L$-least counterexample. Then X is definable (without parameters) in L_{κ^+}, since $\langle N_a\rangle$ is definable in L_{κ^+}. Choose $\tau \in \bar{S}_\kappa$ s.t. $L_\tau \prec L_{\kappa^+}$. We shall show

Claim 1: If $a \in \mathfrak{P}_\kappa(\kappa)$ and $\Theta = \Theta(a)$ is a limit point of $\bar{B}_\tau \cap a$, then $X \cap \Theta$, $\bar{B}_\tau \cap \Theta \in N_a$

This contradicts the definition of X.
So let a be given and set $\Theta = \Theta(a)$. Since Θ is a limit point of $\bar{B}_\tau \cap a$ there is some $\bar{\tau} \in \bar{S}_\Theta$ s.t. $\bar{\tau} \prec_\kappa \tau$. We only have to show

Claim 2: $\bar{\tau} \in N_a$

To see this, first observe that $\bar{B}_\tau \cap \Theta = \bar{B}_{\bar{\tau}}$ and N_a is closed under $\eta \longrightarrow \bar{B}_\eta$. We also know that $X \cap \Theta$ is definable in $L_{\bar{\tau}}$.

We now prove Claim 2. We have to show that $\bar{\tau} \in \mathrm{rng}(f_a)$. Set $f = f_a$ and let $f : \bar{\nu} \Longrightarrow \nu$. Let $\langle \eta(\delta) | \delta < \rho\rangle$ be the increasing enumeration of $\bar{B}_\tau \cap a$. By definition, $\{\eta(\delta) | \delta < \rho\} \subseteq \mathrm{rng}(f)$. So let $f(\bar{\eta}(\delta)) = \eta(\delta)$ for $\delta < \rho$. Now $\langle \eta(\delta) | \delta < \rho\rangle$ is a convergent $\prec$-chain. Hence $\langle \bar{\eta}(\delta) | \delta < \rho\rangle$ is a convergent $\prec$-chain by (E5). So let $\bar{\eta}$ be the minimal $\prec$-successor of the sequence $\langle \bar{\eta}(\delta)\rangle$ and let $\bar{\Theta} = \alpha_{\bar{\eta}}$. Obviously, $f(\bar{\Theta}) = \Theta$. But then $f(\bar{\eta})$ is a successor of $\langle \eta_\delta\rangle$ by (E 2), (E 3) and $f(\bar{\eta}) \in \bar{S}_\Theta$. Hence $f(\bar{\eta}) = \bar{\tau}$. qued

The same method can be used to prove $\diamondsuit^+_{\kappa,\lambda}$ for any regular κ, if $\lambda<\kappa$. We now turn to Theorem 1 which was stated at the beginning of this chapter.

Proof of Theorem 1:

Let κ be given. We shall show that $\underset{\sim}{T}^\kappa$ is a rich κ-tree, where $\underset{\sim}{T}^\kappa = \langle T^\kappa, \subseteq\rangle$ is the tree given by the family $F_\kappa = \{\bar{B}_\nu | \nu \in \bar{S}_\kappa\}$. Clearly, $|F_\kappa| = \kappa^+$. Now observe that in the proof of Lemma 2 we implicitly showed:

Let $\tau \in \bar{S}_\kappa$. If $a \in \mathfrak{B}_\kappa(\kappa)$ and $\Theta = \Theta(a)$ is a limit point of $\bar{B}_\tau \cap a$, then $\bar{B}_\tau \cap \Theta \in N_a$. So F_κ satisfies (ii) in Def. 1. So we only have to show (iii). For $\lambda=\kappa$ we proved this in § 1. So let $\lambda<\kappa$. It is easy to see that it suffices to show the corresponding statement for the tree $\langle T', \prec_\kappa\rangle$ where $T' = \bigcup_{\nu\in \bar{S}_\kappa} \bar{B}_\nu$. So let $\bar{T} \subseteq T'$ s.t. $\langle \bar{T}, \prec_\kappa\rangle$ is a λ-tree. We have to show that $\bar{T}$ has a branch of length λ. Let $\rho = \min \{\rho| \, |\bar{T}\cap\rho| = \lambda\}$. Clearly, $cf(\rho)=\lambda$. We may assume w.l.o.g. that $\bar{T}\subseteq\rho$.

Case 1: $\lambda<\rho$, hence ρ is singular.

Choose $\tau \in S_\kappa$ s.t. $\bar{T} \in L_\tau \prec L_{\kappa^+}$. For $\alpha<\lambda$ let $f_\alpha = f_{(\alpha,T,\tau)}$. Since $\lambda>\omega$ is regular, there is some $\alpha<\lambda$ s.t. $\lambda \cap rng(f_\alpha) = \alpha$. Set $f_\alpha = f$ and let $f : \bar{\tau} \Longrightarrow \tau$. Then $f(\alpha) = \lambda$ and $f\restriction\alpha = id\restriction\alpha$. Let $f(\tilde{T}) = \bar{T}$. Since f is elementary and $\bar{T}$ is a λ-tree, we get $f''\tilde{T} = \bigcup_{\beta<\alpha} \bar{T}_\beta$. But there is a convergent $\prec$-chain $\langle\eta(\beta)|\beta<\alpha\rangle$ s.t. $\eta(\beta)\in \bar{T}_\beta$. So let $f(\bar{\eta}(\beta)) = \eta(\beta)$ for $\beta<\alpha$. Then $\langle\bar{\eta}(\beta)|\beta<\alpha\rangle$ is a convergent $\prec$-chain by (E 5). Let $\bar{\eta}$ be the minimal $\prec$-successor of $\langle\bar{\eta}(\beta)\rangle$. Clearly, $\bar{\eta} \in S_{\bar{\rho}}$ where $f(\bar{\rho}) = \rho$ and $L_{\bar{\eta}} \models$ "$\bar{\rho}$ is regular". Hence $\bar{\eta}<\bar{\tau}$, since $L_{\bar{\tau}} \models$ "$\bar{\rho}$ is singular". So let $\eta = f(\bar{\eta})$. Then $B_\eta \cap \bar{T}$ is a "cofinal" branch of $\bar{T}$ (of cardinality λ).

Case 2: $\lambda=\rho$

In this case, the argument is essentially the same as in the proof of Lemma 4, § 1, so we only indicate the proof. Choose $\nu\in S_\lambda$ s.t. $\bar{T}, \vec{A}|\lambda\in L_\nu$, where $\vec{A} = \langle A_\alpha|\alpha<\kappa\rangle$ is the sequence we used to define $\prec_\kappa$. Let $\bar{\nu}\prec\nu$ be minimal s.t. $\bar{T}, \vec{A}\restriction\lambda\in \text{rng}\,\Pi_{\bar{\nu}\nu}$. Then $B_{\bar{\eta}}\cap\tilde{T}$ is "cofinal" in $\tilde{T}$ for some $\bar{\eta}\in S_{\bar{\alpha}}\cap\bar{\nu}$, where $\bar{\alpha}=\alpha_{\bar{\nu}}$ and $\Pi_{\bar{\nu}\nu}(\tilde{T}) = \bar{T}$. Hence $B_\eta\cap\bar{T}$ is "cofinal" in $\bar{T}$ where $\eta = \Pi_{\nu\nu}(\bar{\eta})$.

qued.

We now give an application of rich κ-trees.

Proposition 3: Assume there is a rich κ-tree. Then there is some $H : [\kappa^+]^2 \longrightarrow \kappa$, which satisfies:

Let $X\subseteq\kappa^+$ and $|X|=\lambda$, where $\omega\leq\lambda\leq\kappa$ and λ is regular. Then there are injective $g : \lambda \longrightarrow \kappa$ and $h : \lambda \longrightarrow X$ s.t. $H(\{h(\nu), h(\tau)\}) = g(\nu)$ for every $\nu<\tau<\lambda$.

Proof: Let $\underset{\sim}{T} = \langle T,\subseteq\rangle$ be a rich κ-tree given by the family $F\subseteq\mathfrak{B}(\kappa)$. It suffices to find $H : [F]^2 \longrightarrow T$ having the analogous property. We define $H : [F]^2 \longrightarrow T$ by $H(\{B,\bar{B}\}) = g_B\restriction\nu$ where $\nu = \max\{\rho<\kappa|B\cap\rho = \bar{B}\cap\rho\}$. Now let $\bar{F}\subseteq F$ be given and let $|\bar{F}| = \lambda$ where $\lambda\leq\kappa$ is regular. Let $\bar{T} = H''[\bar{F}]^2$. It suffices to show:

Claim: $\bar{T}$ has a branch of length λ

For then let $g : \lambda \longrightarrow \bar{T}$ be the increasing enumeration of such a branch. Define $h : \lambda \longrightarrow \bar{F}$ so that for every $\nu<\lambda$ there is some $\bar{B}\in \bar{F}$ s.t. $H(\{h(\nu), \bar{B}\}) = g(\nu)$. Clearly, $H(\{h(\nu), h(\tau)\}) = g(\nu)$ whenever $\nu<\tau<\lambda$.

We now prove the claim. We show by induction that $|\bar{T}_\alpha|<\lambda$ for all $\alpha<\lambda$. Then the claim follows from condition (iii) for the rich κ-tree $\underset{\sim}{T}$. So let $\alpha<\lambda$. Set $D = \{\text{dom}\, f|f\in \bar{T}_\beta$ for some $\beta<\alpha\}$, hence $|D|<\lambda$ by induction hypothesis. Hence $|F\restriction D|<\lambda$ by condition (ii) for $\underset{\sim}{T}$. But simple tree arguments show that $g\restriction D \neq \bar{g}\restriction D$, if $g,\bar{g}\in \bar{T}_\alpha$, $g\neq\bar{g}$. Hence $|\bar{T}_\alpha|<\lambda$.

qued.

Corollary: Assume there is a rich κ-tree. Then there are $H_n : [\kappa^+]^{n+1} \longrightarrow [\kappa]^n (1 \le n < \omega)$ s.t.: for every $X \subseteq \kappa^+$ s.t. $\omega \le |X| \le \kappa$ and $|X|$ is regular there is some $Y \subseteq \kappa$ s.t. $|X| = |Y|$ and $H_n'' [X]^{n+1} \supseteq [Y]^n$

Proof: Let H be defined as above and set $H_n(\{\nu_0,\ldots,\nu_n\}) = \{H(\{\nu_i,\nu_j\}) \mid i<j\le n\}$.

qued.

This result gives a rather general stepping-up lemma in L for various kinds of negative partition relations. For example, recall the definition of the square-bracket relation.

Definition: $\kappa \longrightarrow [\lambda_\nu]^n_\tau$ iff for all $f[\kappa]^n \longrightarrow \tau$ there is some $X \subseteq \kappa$ and $\nu < \tau$ s.t. $|X| = \lambda_\nu$ and $\nu \notin f''[X]^n$.

The following result gives a partial answer to Problem 17 in [4] under the assumption V=L. Todorevic has proved the consistency of the statement (see [7]).

Proposition 4: Assume V=L. Let $\kappa > \omega$ be regular and $\tau \le \kappa$, $\omega \le \lambda_\nu \le \kappa$, λ_ν regular, $n \ge 1$. Then:

$$\kappa \not\longrightarrow [\lambda_\nu]^n_\tau \curvearrowright \kappa^+ \not\longrightarrow [\lambda_\nu]^{n+1}_\tau$$

Proof: We may assume that κ is not ineffable, since otherwise the left side is false. So let $H_n : [\kappa^+]^{n+1} \longrightarrow [\kappa]^n$ be as in the corollary. Let $f : [\kappa]^n \longrightarrow \tau$ give the relation on the left side. Then $f \circ H_n$ gives the relation on the right side.

qued.

Note that we could also treat finite λ_ν's above. Finally, let $<_*$ be the lexicographical ordering on ${}^\kappa 2$. For $\kappa = \omega_1$ the following result was proved by Devlin in [2].

Proposition 5: Let $F \subseteq \mathfrak{P}(\kappa)$ be a family s.t. the tree given by F ir rich. Let $M = \{g_B | B \in F\}$. Then given any $X \subseteq M$ s.t. $|X| \le \kappa$, $|X|$

regular, there is some $Y \subseteq X$ s.t. $|Y| = |X|$ and Y is wellordered by $<_*$ or $>_*$.

(Recall that there is no $X \subseteq {}^{\kappa}2$ s.t. $|X| = \kappa^+$ and X is wellordered by $<_*$ or $>_*$).

<u>Proof</u>: This follows immediately from the proof of Proposition 3.

qued.

<u>References</u>

[1] K.J. Devlin, Aspects of constructibility, Springer Lecture Notes in Mathematics 354 (1973)

[2] K.J. Devlin, Order-types, trees, and a problem of Erdős and Hajnal Periodica Math. Hungarica 5 (1974), pp.153-160

[3] K.J. Devlin, The combinatorial principle $\diamondsuit^{\#}$, to appear

[4] P. Erdős and A. Hajnal, Unsolved problems in set theory, in: Axiomatic Set Theory, Proc.Symp.Pure Math.Vol.23, Part I(1971) pp. 17-48

[5] R.B. Jensen, The (κ,β)-morass (unpublished manuscript)

[6] L.J. Stanley, "L-like" models of set theory: forcing, combinatorial principles and morasses, Thesis, Berkeley (1977)

[7] St.B. Todorevic, Some results in set theory II, Notices of the AMS (1979), A 440

[8] K. Wolfsdorf, Der Beweis eines Satzes von G. Choodnovsky, Arch.Math.Logik 20 (1980), pp. 161-171

[9] F.G. Abramson, L.A. Harrington, E.M. Kleinberg, W.S. Zwicker, Flipping properties: a unifying thread in the theory of large cardinals, Ann.of Math.Logic 12 (1977), pp. 25-58

SOME APPLICATIONS OF THE CORE MODEL

D. Donder, Bonn, R.B. Jensen, Oxford, and B.J. Koppelberg, Berlin.

In § 1, § 2 and § 4 of this paper we apply the core model K to questions about partition properties of cardinals and ultrafilters. In § 3 we prove an analogue of Schoenfield's absoluteness theorem for K.

In [10] Mitchel proved that Ramsey cardinals are absolute in K and that, if κ is Ramsey, then there is a smallest inner model in which it has the property. In § 1 we improve Mitchell's result to:

<u>Theorem 1</u> (Jensen): Let κ be α-Erdös, where $cf(\alpha) > \omega$. Then κ is α-Erdös in K.

Following Mitchell, we then get:

<u>Corollary 2</u>: Let κ be α-Erdös, where $cf(\alpha) > \omega$. There is a smallest inner model in which it has the property.

We then use own methods to determine the exact strength of Chang's conjecture. Silver showed that Chang's conjecture is consistent relative to the existence of an ω_1-Erdös cardinal. We show the converse:

Theorem 3 (Donder): Assume Chang's conjecture. Let $\kappa = \omega_2$, $\alpha = \omega_1$. Then κ is α-Erdös in K.

In § 2 we address ourselves to the existence of non regular and weakly normal ultrafilters. Let $\neg L^{\mu}$ be the statement that there is no inner model with a measurable cardinal.

Theorem 4 (Jensen): Assume $\neg L^{\mu}$. Let κ be an infinite cardinal. Then every uniform ultrafilter on $\kappa^{+}$is (κ,κ^{+})-regular.

Ketonen in [8] proved this result on the assumption $\neg 0^{\#}$. Our proof is based on his method.

Theorem 5 (Jensen): Assume $\neg L^{\mu}$. Let κ be regular such that $2^{\underline{\kappa}} = \kappa$. Then there is no weakly normal ultrafilter on κ.

R.B. Jensen and B.J. Koppelberg originally proved this result on the assumption $\neg 0^{\#}$, in the wake of Ketonen's work on (κ,κ^{+})-regularity (Since this paper was written, Donder has vastly extended these results, proving under the assumption of V = K that all uniform ultrafilters are regular).

The main theorem of § 3 reads

Theorem 6 (Jensen): If $0^{\dagger}$ does not exist but $a^{\#}$ exists for every real $a^{\#}$, then K is Σ^1_3-absolute.

As a corollary of the proof we obtain:

Theorem 7 (Jensen): Let A be Π^1_2. Let M be a mouse. Assume A(a), where $M \notin L[a]$. Then $\exists a \in L[M](A(a))$.

§ 4 contains Koppelberg's work on decomposability.

In [12] two results on decomposability are presented:

1) (Jensen, K. Prikry and J. Silver)

Assume $V = L$. Let $\kappa \geq \omega$ be regular and not weakly compact. Let U be uniform over κ. Then U is λ-decomposable for all $\lambda \leq \kappa$. (Proposition 1 and theorem 20).

2) (J. Silver) Suppose that for some strongly inaccessible cardinal κ there is an ultrafilter over κ which is λ-indecomposable for all λ s.th. $\omega < \lambda < \kappa$. Then $0^{\#}$ exists.

These results generalize to the following two theorems:

Theorem 4.3. (B. Koppelberg) Assume $\neg L^{\mu}$ and let U be a uniform ultrafilter on some regular cardinal κ. Then U is δ-decomposable for all regular cardinals $\delta \leq \kappa$.

Let $\mu \leq \kappa$ be cardinals and let $LCH_{\mu,\kappa}$ be the statement, that all limit cardinals $\lambda < \kappa$ s.th. $\mu \leq cf\lambda < \lambda$ are strong limit cardinals.

Theorem 4.7. (B. Koppelberg) Assume $\neg L^{\mu}$ and let U be a uniform ultrafilter on some regular cardinal κ. Assume $LCH_{\omega_2,\kappa}$. Then U is δ-decomposable for all $\delta \leq \kappa$.

The methods leading to these results left the case of uniform ultrafilters on singular cardinals impregnable. But since the paper has been written, D. Donder has obtained a result for this case analogous to theorem 4.3.

§ 1. Partition Cardinals in K

In [1] Baumgartner introduces the following useful definition.

Let $\omega \leq \alpha \leq \kappa$. κ is α-Erdös iff whenever C is a closed unbounded subset of κ and $f:[C]^{<\omega} \longrightarrow \kappa$ is regressive (i.e. $f(\vec{v}) < \min(\vec{v})$) there is a $D \subseteq C$ such that $|D| \leq \alpha$ and D is homogeneous for f.

(Concerning our notation let us mention that, if A is a set of ordinals, $|A|$ denotes the order type of A. On the other hand, $\overline{\overline{X}}$ denotes the cardinality of X).

We refer the reader to Baumgartner's paper for some facts about α-Erdös cardinals. Being α-Erdös is obviously a strengthening of the partition relation $\kappa \longrightarrow (\alpha)^{<\omega}$.
Conversely, if α is a limit ordinal and κ is the least κ such that $\kappa \longrightarrow (\alpha)^{<\omega}$, then κ is α-Erdös. So especially κ is Ramsey iff κ is κ-Erdös. It is a pleasant property of α-Erdös cardinals that they are subtle and therefore inaccessible.

In [10] Mitchell showed that Ramsey cardinals are Ramsey in the core model K. We improve this to:

Theorem 1.1: Let κ be α-Erdös and $cf(\alpha) > \omega$. Then κ is α-Erdös in K.

The following example shows that some assumption about α is necessary. Let κ be ω_1-Erdös and set $\alpha = \omega_1^L$. We construct an inner model M such that κ is α-Erdös in M but not in K^M. Since $\mathfrak{P}^L(\alpha)$ is countable there is an L-generic collapsing map $f:\omega \longrightarrow \alpha$. Let $M = L[f]$. Since α is countable in M a well-known argument due to

Silver yields that κ is α-Erdös in M. But $K^M = L$ because M is a generic extension of L. Therefore κ cannot be α-Erdös in K^M.

It is well-known that the partition relation $\kappa \longrightarrow (\alpha)^{<\omega}$ can be reformulated in terms of indiscernibles for structures. We show now that this can also be done for α-Erdös cardinals.

Definition. Let $A_1,\ldots,A_n \subseteq \kappa$ and set $\mathfrak{A} = \langle L_\kappa[\vec{A}], \in, \vec{A}\rangle$, $\mathfrak{A}_\beta = \mathfrak{A} | L_\beta[\vec{A}]$ for $\beta < \kappa$.
$I \subseteq \kappa$ is a good set of indiscernibles for $\mathfrak{A}$ (or good for $\mathfrak{A}$) iff for all $\gamma \in I$:

(G 1) $\mathfrak{A}_\gamma \prec \mathfrak{A}$

(G 2) $I-\gamma$ is set of indiscernibles for $\langle \mathfrak{A}, (\xi)_{\xi<\gamma}\rangle$

Lemma 1.2: Let $\alpha \geq \omega$ be a limit ordinal. κ is α-Erdös iff every model $\mathfrak{A} = \langle L_\kappa[\vec{A}], \in, \vec{A}\rangle$ has a good set of indiscernibles of order type α.

Proof: Let κ be α-Erdös. Given $\mathfrak{A}$ as above set $C = \{\gamma < \kappa \,|\, \mathfrak{A}_\gamma \prec \mathfrak{A}\}$. Then C is closed unbounded in κ. Let $\langle \varphi_\beta \mid \beta < \kappa\rangle$ be an enumeration of the formulas for $\langle \mathfrak{A}, (\xi)_{\xi<\kappa}\rangle$ such that for $\gamma \in C$ formulas with parameters from γ have numbers less than γ. Define $f:[C]^{<\omega} \longrightarrow \kappa$ such that $f(\{\vec{\nu},\vec{\mu}\}) = 0$, if $\vec{\nu},\vec{\mu}$ realize the same type over $\langle \mathfrak{A}, (\xi)_{\xi<\nu}\rangle$ and otherwise $f(\{\vec{\nu},\vec{\mu}\})$ is the number of a formula which gives a counterexample. Let $I \subseteq C$ be homogeneous for f such that $|I| = \alpha$. Then I is good for $\mathfrak{A}$.

The opposite direction is obvious.

It is now clear that the following indiscernibility lemma is a strengthening of Theorem 1.1.

Lemma 1.3: Let $A \subseteq \kappa$ such that $L_\kappa[A] \subseteq K_\kappa$. Set $\mathfrak{A} = \langle K_\kappa, \in, D \cap \kappa, A\rangle$. Let I be good for $\mathfrak{A}$ such that $cf(|I|) > \omega$. Then there is $I' \in K$ such that that I' is good for $\mathfrak{A}$ and $I \subseteq I'$.

Proof: By standard indiscernibility arguments each $\gamma \in I$ is inaccessible in $\mathfrak{A}$. Since $\mathfrak{A}_\gamma \prec \mathfrak{A}$ it follows that $\mathfrak{A}$ is a model of ZFC.

For $\beta \in \mathfrak{A}$ set $\beta^+ = (\beta^+)^{\mathfrak{A}}$.

We define models $\bar{\mathfrak{A}}^n_\gamma \prec \mathfrak{A}_{\gamma^+}$ for $\gamma \in I$ and $n < \omega$ as follows.

$\bar{\mathfrak{A}}^n_\gamma$ = the restriction of $\mathfrak{A}$ to the set of $x \in \mathfrak{A}_{\gamma^+}$ which are $\mathfrak{A}$-definable from parameters in $\gamma \cup \{\gamma, \gamma_1, \ldots, \gamma_n\}$ where $\gamma < \gamma_1 < \ldots < \gamma_n$, $\gamma_i \in I$.

Clearly, $\bar{\mathfrak{A}}^n_\gamma$ is transitive, so $\bar{\mathfrak{A}}^n_\gamma = \mathfrak{A}_{\delta(\gamma,n)}$ for a $\delta(\gamma,n) < \gamma^+$. Using (G1) we see that we can replace "$\mathfrak{A}$-definable" by "$\mathfrak{A}_{\gamma_{n+1}}$-definable" in the definition above where $\gamma_{n+1} \in I$, $\gamma_{n+1} > \gamma_n$. This yields

(1) $\bar{\mathfrak{A}}^n_\gamma \in \mathfrak{A}^{n+1}_\gamma$

Set $\mathfrak{A}_\gamma = \bigcup_{n<\omega} \bar{\mathfrak{A}}^n_\gamma$. It is obvious that

(2) $\bar{\mathfrak{A}}^n_\gamma \prec \bar{\mathfrak{A}}^{n+1}_\gamma \prec \bar{\mathfrak{A}}_\gamma \prec \mathfrak{A}_{\gamma^+}$

Let $h^{\mathfrak{A}}_i$ $(i \in \omega)$ be a complete set of definable Skolem functions for $\mathfrak{A}$. Since I satisfies (G2) we can define elementary embeddings $\pi^n_{\gamma\gamma'} : \bar{\mathfrak{A}}^n_\gamma \longrightarrow \bar{\mathfrak{A}}^n_{\gamma'}$ for $\gamma \leq \gamma'$ by:

$$\pi^n_{\gamma\gamma'}(h^{\mathfrak{A}}_i(\vec{\nu}, \gamma, \gamma_1, \ldots, \gamma_n)) = h^{\mathfrak{A}}_i(\vec{\nu}, \gamma', \gamma_1, \ldots, \gamma_n)$$

where $\vec{\nu} < \gamma \leq \gamma' < \gamma_1, \ldots, \gamma_n$, $\gamma_j \in I$

Arguing as for (1) we see that $\pi^n_{\gamma\gamma'} \in \mathfrak{A}$ and is $\mathfrak{A}$-definable in n+3 parameters from I. Now set $\pi_{\gamma\gamma'} = \bigcup_{n\in\omega} \pi^n_{\gamma\gamma'}$. Using (2) we get

(3) $\pi_{\gamma\gamma'} : \bar{\mathfrak{A}}_\gamma \xrightarrow[\Sigma\omega]{} \bar{\mathfrak{A}}_{\gamma'}$

It is clear that $\pi_{\gamma\gamma'} \upharpoonright \gamma = \mathrm{id} \upharpoonright \gamma$ and $\pi_{\gamma\gamma'}(\gamma) = \gamma'$. Now define $U_\gamma \subseteq \mathfrak{P}(\gamma) \cap \bar{\mathfrak{A}}_\gamma$ by:

$$X \in U_\gamma \longleftrightarrow \gamma \in \pi_{\gamma\gamma'}(X) \quad \text{for} \quad \gamma < \gamma'$$

(4) U_γ is normal on γ in $\langle \bar{\mathfrak{A}}_\gamma, U_\gamma \rangle$
and $\langle \bar{\mathfrak{A}}_\gamma, U_\gamma \rangle$ is amenable .

Normality is obvious. To prove amenability set $U^n_\gamma = U_\gamma \cap \bar{\mathfrak{A}}^n_\gamma$. It suffices to show that $U^n_\gamma \in \bar{\mathfrak{A}}_\gamma$. But this is immediate since U^n_γ is $\mathfrak{A}$-definable from $\pi_{\gamma\gamma'}$ and $U^n_\gamma \in \mathfrak{A}_{\gamma+}$.

Since the definition of U^n_γ is uniform we get $\pi_{\gamma\gamma'}(U^n_\gamma) = U^n_{\gamma'}$. So observing that $\pi_{\gamma\gamma'}$ is cofinal we have:

(5) $\pi_{\gamma\gamma'} : \langle \bar{\mathfrak{A}}_\gamma, U_\gamma \rangle \xrightarrow[\Sigma_1]{} \langle \bar{\mathfrak{A}}_{\gamma'}, U_{\gamma'} \rangle$

Now set $\gamma^* = \sup I$, $I^* = I \cup \{\gamma^*\}$. Clearly, $\mathfrak{A}_{\gamma*} \prec \mathfrak{A}$. Let $\langle \bar{\mathfrak{A}}_{\gamma*}, U_{\gamma*} \rangle$, $\langle \pi_{\gamma\gamma*} \mid \gamma \in I \rangle$ be the direct limit of $\langle\langle \mathfrak{A}_\gamma, U_\gamma \rangle \mid \gamma \in I \rangle$, $\langle \pi_{\gamma\gamma'} \mid \gamma \le \gamma' , \gamma,\gamma' \in I \rangle$. $\bar{\mathfrak{A}}_{\gamma*}$ is well founded, since $\mathrm{cf}(|I|) > \omega$, and may therefore be taken as transitive. It is clear that (3) - (5) go through for $\gamma,\gamma' \in I^*$.

Let $M^\gamma = J^{U_\gamma}_{\beta_\gamma} = \bigcup\{ J^{U_\gamma}_{\nu+1} \mid J^{U_\gamma}_\nu \in \bar{\mathfrak{A}}_\gamma \}$ for $\gamma \in I^*$. Since $\langle \bar{\mathfrak{A}}_\gamma, U_\gamma \rangle$ is rud closed we see that $J^{U_\gamma}_\nu \in \bar{\mathfrak{A}}_\gamma$ implies $J^{U_\gamma}_{\nu+1} \subseteq \mathfrak{A}_\gamma$. So M^γ is a premouse and we know $M^\gamma \subseteq \bar{\mathfrak{A}}_\gamma$, $M^\gamma \notin \bar{\mathfrak{A}}_\gamma$. We now show:

(6) M^γ is iterable.

Noting that $\pi_{\gamma\gamma'} \upharpoonright M^\gamma : M^\gamma \xrightarrow[\Sigma_0]{} M^{\gamma^*}$ it suffices to show that M^{γ^*} is iterable. This in turn will follow from the fact that $U_{\gamma*}$ is ω-complete. But this is clear since $\mathrm{cf}(\gamma^*) > \omega$ and:

$$X \in U_{\gamma^*} \longleftrightarrow X \in \bar{\mathfrak{A}}_\gamma \text{ and } \exists \gamma < \gamma^* \; I - \gamma \subseteq X$$

Our next aim is to show that $\mathfrak{A}_\gamma \in M^\gamma$. We need first:

$$(7) \; \mathfrak{P}(\gamma) \cap M^\gamma \not\subseteq \bar{\mathfrak{A}}^n_\gamma \quad \text{for} \quad \gamma \in I, n < \omega .$$

Suppose not and set $\beta = \beta_\gamma$, $\bar{U} = U^n_\gamma$. Then $M^\gamma = J^{\bar{U}}_\beta$. Since $\mathfrak{A}$ is a model of ZFC satisfying $V = K$, $\bar{U} \in \mathfrak{A}$ is not normal in $L^{\bar{U}}_\kappa$. Hence there is a least $\tau < \kappa$ such that $\mathfrak{P}(\gamma) \cap J^{\bar{U}}_{\tau+1} \not\subseteq \mathfrak{A}^n_\gamma$. Clearly, $\beta \leq \tau < \gamma^+$. But then $\tau \in \bar{\mathfrak{A}}_\gamma$, since $\bar{\mathfrak{A}}_\gamma < \mathfrak{A}_{\gamma+}$. Hence $\beta \in \bar{\mathfrak{A}}_\gamma$ and $M^\gamma = J^{\bar{U}}_\beta \in \bar{\mathfrak{A}}_\gamma$, contradicting an earlier observation.

$$(8) \; \mathfrak{P}(\gamma) \cap \bar{\mathfrak{A}}_\gamma \subseteq \mathfrak{P}(\gamma) \cap M^\gamma \quad \text{for} \quad \gamma \in I^*$$

It suffices to show this for $\gamma \in I$. So let $a \in \mathfrak{P}(\gamma) \cap \bar{\mathfrak{A}}_\gamma$. Since $\mathfrak{P}(\gamma) \cap L \subseteq M^\gamma$ by iterability, we may assume $a \notin L$. But then $\mathfrak{A}$ satisfies ZFC + $V = K + V \neq L$. Moreover the notion of mouse is absolute in $\mathfrak{A}$, since $\omega_1 \leq \gamma^* \leq \kappa$. Hence there is a mouse $N \in \mathfrak{A}_{\gamma+}$ at a $\tau \geq \gamma$ such that $a \in N$. Since $\bar{\mathfrak{A}}_\gamma \prec \mathfrak{A}_{\gamma+}$ we may assume $N \in \bar{\mathfrak{A}}_\gamma$. Let N_θ be the mouse iteration of $\mathbb{N}$ to a regular $\theta > \kappa$ and let M_θ be the iteration of M^γ to θ. Then $M_\theta \subseteq N_\theta$ or $N_\theta \subseteq M_\theta$. If $N_\theta \subseteq M_\theta$, we get $a \in M$. So assume $M_\theta \subseteq N_\theta$ and choose n such that $N \in \bar{\mathfrak{A}}^n_\gamma$. Then $\mathfrak{P}(\gamma) \cap M^\gamma = \mathfrak{P}(\gamma) \cap M_\theta \subseteq \mathfrak{P}(\gamma) \cap N_\theta \subseteq N \subseteq \bar{\mathfrak{A}}^n_\gamma$, which contradicts (7). This proves (8).

Eventually, we are ready to finish the proof of Lemma 1.3. Using (8) let ν_γ be the least $\nu < \beta_\gamma$ such that $\mathfrak{A}_\gamma \in J^{U_\gamma}_{\nu+1}$. Clearly, $\pi_{\gamma\gamma'}(\mathfrak{A}_\gamma) = \mathfrak{A}_{\gamma'}$, hence $\pi_{\gamma\gamma'}(\nu_\gamma) = \nu_{\gamma'}$. Set $N^\gamma = J^{U_\gamma}_{\nu_\gamma+1}$. Then, since $\mathfrak{P}(\gamma) \cap J^{U_\gamma}_{\nu+1} \not\subseteq J^{U_\gamma}_\nu$, there is a $\Sigma_1(N_\gamma)$ map of a subset of γ onto N^γ. Hence N^γ is a mouse, since it is iterable. Let $\langle \varphi_i \mid i < \omega \rangle$ be a recursive enumeration of the first order formulas for $\mathfrak{A}$. By standard methods (cf. the proof of

Lemma 1.2), one may canonically choose $X^i_\gamma \in N^\gamma$ such that $X^i_\gamma \in U_\gamma$ and if $\nu \in X^i_\gamma$, $\vec{\xi} < \nu$ and $<\vec{\tau}>$, $<\vec{\tau}'> \in [X^i_\gamma - \nu]^n$, then $\mathfrak{A}_\gamma \models \varphi_i(\vec{\xi},\vec{\tau}) \longleftrightarrow \mathfrak{A}_\gamma \models \varphi_i(\vec{\xi},\vec{\tau}')$.

By the canonical choice we have $\pi_{\gamma\gamma'}(X^i_\gamma) = X^i_{\gamma'}$. In particular, $\gamma \in X^i_{\gamma'}$ for $\gamma < \gamma'$.
Now set $I' = \bigcap_{i<\omega} X^i_{\gamma^*}$. Then $I \subseteq I'$ and I' is good for $\mathfrak{A}_{\gamma^*}$, hence for $\mathfrak{A}$, since $\mathfrak{A}_{\gamma^*} \prec \mathfrak{A}$. But I' was defined from N^{γ^*} and $N^{\gamma^*} \in K$, since N^{γ^*} is a mouse. So $I' \in K$ is the set we were looking for.

Before proceeding further let us remind the reader that there is a natural prewellordering of the mice which is defined as follows

$$M \lhd N \quad \text{iff} \quad M_\theta \in N_\theta \quad \text{where} \quad \theta > \overline{\overline{M}},\overline{\overline{N}} \text{ is regular.}$$

It is shown in [4] that $\lhd$ restricted to the core mice is a well-ordering.
We often use the following simple fact.

Let $M \lhd N$, where M,N are mice at κ,τ respectively. Then $\mathcal{P}(\gamma) \cap M \subseteq N$ for every $\gamma \leq \kappa,\tau$. To prove this let $\theta > \kappa,\tau$ be regular. Then $\mathcal{P}(\gamma) \cap M = \mathcal{P}(\gamma) \cap M_\theta \subseteq \mathcal{P}(\gamma) \cap N_\theta \subseteq N$.

Our next aim is to describe all inner models which satisfy $V = K$. Call a mouse M at κ <u>critical</u> iff $H^M_\kappa \in M$. It is easily seen (cf. the proof of Lemma 4.9 in [4]) that, if M is critical, then $H^M_\kappa = K^M_\kappa$ and $H^M_\kappa \models ZF$. Moreover, if M_i is an iterative of M at κ_i, then $H^{M_i}_{\kappa_i} \in M_i$ and $H^M_\kappa \prec H^{M_i}_{\kappa_i}$. Hence $\bigcup_{i \in On} H^{M_i}_{\kappa_i}$ is an inner model of $ZFC + V = K$.

<u>Lemma 1.4</u>. Let W be an inner model of ZF. Suppose $K^W \neq K$. Let N

be the $\triangleleft$-least core mouse such that $N \notin W$.
Then N is critical and $K^W = \bigcup_{i \in On} H_{\kappa_i}^{N_i}$, where N_i are the iterates of N at κ_i.

Proof: We first show:

(1) $\mathcal{P}(\kappa_i) \cap K^W \subseteq N_i$

This is clear, if $K^W = L$. So assume $K^W \neq L$ and let $A \in \mathcal{P}(\kappa_i) \cap K^W$. Then there is a mouse $M \in K^W$ at a $\tau \geq \kappa_i$ such that $A \in M$. Since $M \triangleleft N$, we get $A \in N_i$.

(2) $H_{\kappa_i}^{N_i} \subseteq K^W$

Let $a \in H_{\kappa_i}^{N_i}$ and let $N_i = J_\alpha^U$. Assume w.l.o.g. $a \subseteq \gamma$ for some $\gamma < \kappa_i$ and $a \notin L$. Then $a \in J_{\nu+1}^U - J_\nu^U$ for some $\kappa_i < \nu < \alpha$. But then $M = J_\gamma$ is a mouse and $a \in \Sigma_\omega(M)$ (see [4], Corollary 5.2.1). But $M \triangleleft N$, hence $M \in K^W$ which implies $a \in K^W$. This proves (2).

It remains to show that N is critical. Let $A \subseteq \kappa$ recursively code the bounded subsets of κ which are in N. It is enough to show that $A \in N$. But by (2) we know that $A \in K^W$. So $A \in N$ follows from (1).

As a corollary of Lemma 1.4, we get the following result of Mitchell.

Lemma 1.5: Let $\varphi(\vec{x})$ be a formula such that $ZF \vdash \forall \vec{\alpha}\ (\varphi(\vec{\alpha}) \longrightarrow \varphi^K(\vec{\alpha}))$. Assume $\varphi(\vec{\alpha})$. Then we can define an inner model $W = W_{\vec{\alpha}}$ (uniformly in $\vec{\alpha}$) such that

(a) $\varphi(\vec{\alpha})$ holds in W

(b) If Q is an inner model and $\varphi(\vec{\alpha})$ holds in Q, then $W = W^Q$

(Hence W is the smallest inner model in which $\varphi(\vec{\alpha})$ holds).

Proof: If there is a critical core mouse N such that $\varphi(\vec{\alpha})$ holds in

$\bigcup_{i\in On} H_{\kappa_i}^{N_i}$, set $W = \bigcup_{i\in On} H_{\kappa_i}^{N_i}$, where N is the $\triangleleft$-least such. Otherwise set W = K. The conclusion is immediate.

<u>Corollary 1.6</u>: Let κ be α-Erdös and $cf(\alpha) > \omega$. Then there is a smallest inner model in which κ is α-Erdös.

We now turn to an application of the indiscernibility lemma. Chang proposed the following strong two cardinals conjecture.

Let $\mathfrak{A} = \langle A, B, \ldots\rangle$ be a structure of countable length with $\bar{\bar{A}} = \omega_2$, $\bar{\bar{B}} = \omega_1$. Then there is $\langle A', B', \ldots\rangle \prec \mathfrak{A}$ such that $\bar{\bar{A}}' = \omega_1$ and $\bar{\bar{B}}' = \omega$.

Silver proved the consistency of Chang's conjecture starting from a model with an ω_1-Erdös cardinal. On the other hand Kunen showed that Chang's conjecture implies the existence of $0^{\#}$. We strengthen this and show that the assumption in Silver's result cannot be weakened.

<u>Theorem 1.7</u>: Assume Chang's conjecture. Let $\kappa = \omega_2$, $\alpha = \omega_1$. Then κ is α-Erdös in K.

<u>Proof</u>: Let $A \in \mathcal{P}(\kappa) \cap K$ and set $\mathfrak{A} = \langle K_\kappa, \in, D\cap\kappa, A\rangle$. It is our aim to find a good set of indiscernibles for $\mathfrak{A}$ having order type α. This will suffice, since there will then be such a set in K by the indiscernibility lemma.

Let $\kappa^+ = (\kappa^+)^K$ and choose $\rho < \kappa^+$ such that $\mathfrak{A} \in K_\rho$ and $K_\rho \prec K_{\kappa^+}$. Then choose a set $E \subseteq \rho$ such that $K_\rho \subseteq L_\rho[E]$ and α, κ are the first two uncountable cardinals in $L_\rho[E]$. Set $\mathfrak{B} = \langle L_\rho[E], E, K_\rho, D_\rho\rangle$. By Chang's conjecture there is $\mathfrak{B}' \prec \mathfrak{B}$ such that $\bar{\bar{\mathfrak{B}}}' = \omega_1$, $\alpha \cap \mathfrak{B}'$ is countable and

$\mathfrak{a} \in \mathfrak{L}'$. Let $b': \bar{\mathfrak{L}} \xrightarrow{\sim} \mathfrak{L}'$ where $\bar{\mathfrak{L}}$ is transitive. Let $\bar{\mathfrak{L}} = \langle L_{\bar{\rho}}[\bar{E}], \bar{E}, \bar{K}, \bar{D} \rangle$ and $b'(\bar{\mathfrak{a}}) = \mathfrak{a}$. Our choice of E provides that $\alpha \cap \mathfrak{L}'$ is transitive and $b'(\alpha) = \kappa$. Set $\bar{\alpha} = \alpha \cap \mathfrak{L}'$. So $\bar{\alpha}$ is the first point moved by b'. Now set $b = b' \upharpoonright \bar{K}$. Clearly, $b: \langle \bar{K}, \bar{D} \rangle \longrightarrow_{\Sigma_\omega} \langle K_\rho, D_\rho \rangle$.
For later use, note that $\bar{\mathfrak{a}} = \langle \bar{K} \cap K_\alpha, \bar{D} \cap \alpha, \bar{A} \rangle$ where $\bar{A} \subseteq \alpha$ and $b: \bar{\mathfrak{a}} \longrightarrow_{\Sigma_\omega} \mathfrak{a}$.

We consider two cases.

<u>Case 1</u>: $\bar{K} = K_{\bar{\rho}}$.
Set $\tilde{\alpha} = \sup b''\alpha$ and $b_0 = b \upharpoonright K_\alpha$. Then $b_0: K_\alpha \longrightarrow_{\Sigma_1} K_{\tilde{\alpha}}$ cofinally. So a theorem of [4] shows that b_0 can be extended to a $\tilde{b}: K \longrightarrow_{\Sigma_1} K$. But the first point moved by $\tilde{b}$ is taken to α, where α is regular. Hence by another theorem of [3], there is then an inner model L[U] with U normal on α. But it is easy to see that $(\alpha^+)^{L[U]} = (\alpha^+)^K < \kappa$. Hence U can be iterated to κ, yielding U_κ with U_κ normal on κ. But then $\mathfrak{a} \in K \subseteq L[U_\kappa]$ certainly has a good set of indiscernibles of order type α.

<u>Case 2</u>: $\bar{K} \neq K_{\bar{\rho}}$.
Then there is a core mouse $M \notin \bar{K}$ such that the ω-th iterate is a mouse at a $\gamma < \bar{\rho}$. We first show that $M \in K_\alpha$. Recall that $b: \bar{K} \longrightarrow_{\Sigma_\omega} K_{\kappa^+}$. So we observe that $N \lhd M$ for every mouse $N \in \bar{K}$, since otherwise $A_{M'} \in \bar{K}$ which would give $M \in \bar{K}$. In addition, $\bar{K}$ is the union of the mice which are elements of $\bar{K}$. Hence $\bar{K} \subseteq M_\theta$ for $\theta > \bar{\rho}$. But the iteration points of M are regular in M_θ, hence in $\bar{K}$. We also know that α is the largest regular cardinal in $\bar{K}$, since $b(\alpha) = \kappa$. So we have $M \in K_\alpha$.

Since $\bar{\mathfrak{a}} \in \bar{K}$, we also get $\bar{\mathfrak{a}} \in M_\theta$ for θ large enough. But then $\bar{\mathfrak{a}} \in M_\alpha$. Now let $C = C_{M_\alpha}$ be the iteration points of M up to α. Then some final segment $\bar{C}$ of C is good for $\bar{\mathfrak{a}}$. But then $b''\bar{C}$ is a good set of indiscernibles for $\mathfrak{a}$ of order type α, since $b: \bar{\mathfrak{a}} \longrightarrow_{\Sigma_\omega} \mathfrak{a}$.

§ 2 Regularity and Normality of Ultrafilters

We first recall some familiar definitions.
Let U be an ultrafilter on a regular cardinal u. By a regularity sequence for U we mean $\langle u_\nu \mid \nu \in X \rangle$ such that $X \subseteq \kappa$, $u_\nu \subseteq \nu$ for $\nu \in X$ and $\{\nu \mid \eta \in u_\nu\} \in U$ for all $\eta < \kappa$. U is called (γ,κ)-regular iff there is such a sequence satisfying $\bar{\bar{u}}_\nu < \gamma$ for $\nu \in X$. U is weakly normal iff U is uniform and every regressive (mod U) function $f\colon \kappa \to \kappa$ is bounded mod U (i.e. $f''Y \subseteq \gamma$ for some $Y \in U$, $\gamma < u$). It is clear that a weakly normal U contains all closed unbounded subsets of κ. In addition, we have the following diagonalisation principle for weakly normal U:

Let $W \in U$ and $X_\zeta \in U$ for $\zeta < \kappa$. Set $\bar{W} = \{\alpha \in W \mid \alpha = \sup\{\zeta < \alpha \mid \alpha \in X_\zeta\}\}$. Then $\bar{W} \in U$.

To see this, define $f\colon W \to \kappa$ by $f(\alpha) = \sup\{\zeta < \alpha \mid \alpha \in X_\zeta\}$. Assuming $\bar{W} \notin U$, f is regressive mod U. Hence there is $\gamma < \kappa$ such that $Y = \{\alpha \mid f(\alpha) < \gamma\} \in U$. But then $Y \cap X_\gamma = \emptyset$ which contradicts $X_\gamma \in U$.

We now state the main theorem of this chapter. Let $\neg L^\mu$ abbreviate the statement "There is no inner model with a measurable cardinal".

<u>Theorem 2.1</u>: Assume $\neg L^\mu$. Let κ be a regular cardinal. Then every weakly normal ultrafilter on κ is (γ,κ)-regular for some $\gamma < \kappa$.

Kanamori and Ketonen showed (see [6]) that if $\kappa \geq \omega$ and there is a uniform non (κ,κ^+)-regular ultrafilter on κ^+, then κ^+ carries a weakly normal ultrafilter with the same property. Hence we get:

<u>Corollary 2.2</u>: Assume $\neg L^\mu$. Let $\kappa \geq \omega$ be a cardinal. Then every uniform ultrafilter on κ^+ is (κ,κ^+)-regular.

In [8] Ketonen proved Theorem 2.1 under the stronger assumption that $0^{\#}$ does not exist. He showed that an irregular weakly normal ultrafilter can be used to find a non-trivial elementary embedding of L into itself. We make extensive use of his ideas to show that we can actually get such an embedding for K. But we also have to consider two entirely new cases arising from the facts that the condensation lemma need not be true for K and we cannot exclude the possibility $(\kappa^+)^K = \kappa^+$.

Proof of Theorem 2.1

Let κ be a regular cardinal carrying a weakly normal ultrafilter U which is not (γ,κ)-regular for any $\gamma<\kappa$. Considering several cases, we shall show that there is an inner model with a measurable cardinal. We first remark that it suffices to find a non-trivial ultrafilter V over $\mathcal{P}(\kappa)\cap K$ such that the ultrapower of K by V (restricted to functions $f\in K$) is wellfounded. For this gives us a non-trivial elementary embedding of K into some transitive class M. But then M = K by Lemma 6.12 of [4]. So by a theorem of [3] there is an inner model with a measurable cardinal.

We now build a system of structures and embeddings which is essential for the whole proof.

For each $\kappa<\nu<\kappa^+$ pick f_γ mapping κ onto ν. Let $A \subseteq \kappa^+$ code the sets $\langle f_\nu \mid \kappa<\nu<\kappa^+\rangle$ and $D\cap\kappa^+$.
Set $E = \{\tau<\kappa^+ \mid \kappa<\tau \text{ and } \langle L_\tau[A], A\cap\tau\rangle \prec \langle L_{\kappa^+}[A], A\rangle\}$. Then E is closed unbounded in κ^+. For $\tau\in E$, $\alpha<\kappa$ set $Q_\tau = \langle L_\tau[A], A\cap\tau, f_\tau\rangle$ and $\tilde{Q}^\tau_\alpha$ = the smallest $Q \prec Q_\tau$ such that $\alpha \subseteq Q$.

Then $C_\tau := \{\alpha < \kappa \mid \alpha = \kappa \cap Q^\tau_\alpha\}$ is closed unbounded in κ and for every $x \in Q_\tau$ $\{\alpha \in C_\tau \mid x \in \tilde{Q}^\tau_\alpha\}$ is a final segment of C_τ.

Define Q^τ_α, $\tilde{\pi}^\tau_\alpha$ for $\tau \in E$, $\alpha \in C_\tau$ by:

$\tilde{\pi}^\tau_\alpha : Q^\tau_\alpha \stackrel{\sim}{\to} \tilde{Q}^\tau_\alpha$, where Q^τ_α is transitive.

Finally set $\tau_\alpha = On \cap Q^\tau_\alpha$, $K^\tau_\alpha = K^{Q^\tau_\alpha}$ and $\pi^\tau_\alpha = \tilde{\pi}^\tau_\alpha \upharpoonright K^\tau_\alpha$. Hence $\pi^\tau_\alpha : K^\tau_\alpha \xrightarrow{\Sigma_\omega} K_{\kappa+}$.

We often use the following

<u>Fact</u>: Let $\alpha \in C_\tau$, $\eta \in E \cap \tau$, $\eta \in \mathrm{rng}\, \pi^\tau_\alpha$. Then $\alpha \in C_\eta$, $Q^\eta_\alpha = (\pi^\tau_\alpha)^{-1}(Q_\eta) \in Q^\tau_\alpha$ and $\tilde{\pi}^\eta_\alpha = \tilde{\pi}^\tau_\alpha \upharpoonright Q^\eta_\alpha$ (hence $K^\eta_\alpha = (\pi^\tau_\alpha)^{-1}(K_\eta) \in K^\tau_\alpha$, $K^\eta_\alpha \prec K^\tau_\alpha$ and $\pi^\eta_\alpha = \pi^\tau_\alpha \upharpoonright K^\eta_\alpha$).

To verify this just note that Q_η is definable in Q_τ from η, hence $Q_\eta \in Q_\tau$. But then $\tilde{Q}^\tau_\alpha \cap Q_\eta \prec Q_\eta$. Since every $Q \prec Q_\eta$ is uniquely determined by $Q \cap \kappa$, we get $\tilde{Q}^\eta_\alpha = \tilde{Q}^\tau_\alpha \cap Q_\eta$. But $\tilde{Q}^\tau_\alpha$ is an end extension of $\tilde{Q}^\tau_\alpha \cap Q_\eta$, so the rest is clear.

The following sublemma is a key element in our proof.

<u>Lemma 2.3</u>: Let $W \in U$, $f \in \prod_{\alpha \in W} = \alpha^+$. Then there is $\tau \in E$ such that $\{\alpha \in W \cap C_\tau \mid f(\alpha) \leq \tau_\alpha\} \in U$.

<u>Proof</u>: Suppose not. For $\alpha \in W$ let h_α inject $f(\alpha)$ into $\bar{\bar{\alpha}}$. For $\tau \in E$ set $Y_\tau = \{\alpha \in W \sim C_\tau \mid \tau_\alpha < f(\alpha)\}$ and define $g_\tau \in \prod_{\alpha \in Y_\tau} \bar{\bar{\alpha}}$ by $g_\tau(\alpha) = h_\alpha(\tau_\alpha)$.
Our earlier <u>Fact</u> shows that the g_τ are pairwise distinct mod U. Hence some g_{τ^*} has κ many predecessors (mod U) $g_{\tau\zeta}$ $(\zeta < \kappa)$. Since g_{τ^*} is regressive, in fact $g_{\tau^*}(\alpha) < \bar{\bar{\alpha}}$, there is a cardinal $\gamma < \kappa$ such that $Y = \{\alpha \mid g_{\tau^*}(\alpha) < \gamma\} \in U$. Choose $\bar{\tau} \in E$ such that $\bar{\tau} > \tau^*, \tau^\zeta$ and set for $\alpha \in Y \cap C_{\bar{\tau}} =: \bar{Y}$:

$u_\alpha = \{\zeta < \alpha \mid \tau^*, \tau^\zeta \in \mathrm{rug}\, \pi^{\bar{\tau}}_\alpha$ and $g_{\tau\zeta}(\alpha) < g_{\tau^*}(\alpha)\}$.

Then $\langle u_\alpha \mid \alpha \in \bar{Y}\rangle$ is a regularity sequence for U and $\bar{\bar{u}}_\alpha < \gamma$ for $\alpha \in \bar{Y}$.
Hence U is (γ, κ)-regular. Contradiction!

We return to the main line of our proof. We note that K_α^τ is a model of $ZF^- + V = K$ and the notion of mouse is absolute in K_α^τ, since $\pi_\alpha^\tau: K_\alpha^\tau \longrightarrow_{\Sigma_\omega} K_{\kappa^+}$. Hence $K_\alpha^\tau \subseteq K_{\tau_\alpha}$.

Case 1: There is $\tau \in E$ such that $\{\alpha \in C_\tau | K_\alpha^\tau \neq K_{\tau_\alpha}\} \in U$.

Let $\bar{\tau}$ be the least such.

Set $W = \{\alpha | K_\alpha^{\bar{\tau}} \neq K_{\bar{\tau}_\alpha}\}$. Then for each $\alpha \in W$ there is a core mouse $M^\alpha \notin K_\alpha^{\bar{\tau}}$ such that the ω-th iterate of M^α is a mouse at a $\gamma < \bar{\tau}_\alpha$. Let M^α be a mouse at γ^α and let M_i^α be the i-th iterate of M^α with iteration point γ_i^α. Then $\gamma_i^\alpha < \bar{\bar{\alpha}}^+$ for $i < \bar{\bar{\alpha}}^+$.

Lemma 2.4: Let $\tau \in E$, $\tau \geq \bar{\tau}$, $\alpha \in W \cap C_\tau$ and $\bar{\tau} \in \text{rng}\, \pi_\alpha^\tau$ if $\tau > \bar{\tau}$. Let $\gamma_i^\alpha \geq \tau_\alpha$. Then $\mathcal{P}(\nu) \cap K_\alpha^\tau \subseteq M_i^\alpha$ for all $\nu < \tau_\alpha$.

Proof: Suppose not and set $M = M^\alpha$, $\gamma = \gamma^\alpha$. Let $a \subseteq \nu$, $\nu < \tau_\alpha$, such that $a \in K_\alpha^\tau - M_i$. We have $K \neq L$, since M exists. Since $\pi_\alpha^\tau: K_\alpha^\tau \longrightarrow_{\Sigma_\omega} K_{\kappa^+}$, it follows that $a \in N$ for a mouse $N \in K_\alpha^\tau$ at a $\delta \geq \nu, \gamma$. Let $\bar{N} = \text{core}\,(N)$. Then $\bar{N} \not\leq M$ since otherwise $a \in \mathcal{P}(\nu) \cap N \subseteq M_i$. Hence $M \triangleleft N$. But then $A_{M'} \in N \subseteq K_\alpha^\tau$, since $A_{M'} \subseteq \gamma$. Hence $M \in K_\alpha^\tau$, since K_α^τ is a model of ZF^-. But then $M \in K_\alpha^{\bar{\tau}}$, since $K_\alpha^{\bar{\tau}} \prec K$ and M is the unique core mouse with an ω-th iterate at $\gamma_\omega^\alpha < \bar{\tau}_\alpha$. This contradicts our choice of M.

Lemma 2.5: There are arbitrarily large $\tau \in E$ such that

$\{\alpha \in W \cap C_\tau | \bar{\tau} \in \text{rng}\, \pi_\alpha^\tau \text{ and } \exists i\ \tau_\alpha = \gamma_i^\alpha\} \in U$.

Proof: Let $\eta < \kappa^+$. We show that there is $\tau > \eta$ with the property. Define $\tau^\zeta \in E$, $f_\zeta \in \prod_{\alpha \in W} \bar{\bar{\alpha}}^+$ by induction on $\zeta < \kappa$ as follows:

τ^ζ = the least τ such that $\tau > \bar{\tau}, \eta$, $\sup_{\nu<\zeta} \tau^\nu$ and

$\{\alpha \in W \cap C_\tau | f_\zeta(\alpha) < \tau_\alpha\} \in U$;

$f_\zeta(\alpha) = \gamma_i^\alpha$ where i = the least i such that

$$\gamma_i^\alpha > \sup\ \{\tau_\alpha^\nu | \nu < \zeta \text{ and } \alpha \in C_{\tau^\nu}\}.$$

Let $\tau = \sup_{\zeta<\kappa} \tau^\zeta$ and set

$$\bar{C} = \{\alpha \in C_\tau | \sup\ (\mathfrak{C}_\alpha^\tau \cap \tau) = \sup_{\zeta<\alpha} \tau^\zeta\}.$$

Then $\bar{C}$ is closed unbounded in κ. Set $Z = W \cap \bar{C}$ and $Z_\zeta = \{\alpha \in Z | \tau^\zeta \in \text{rng}\ \pi_\alpha^\tau \text{ and } f_\zeta(\alpha) < \tau_\alpha^\zeta\}$ for $\zeta<\kappa$. Then $Z, Z_\zeta \in U$, hence $\bar{Z} = \{\alpha \in Z | \alpha = \sup\ \{\zeta < \alpha | \alpha \in Z_\zeta\}\} \in U$. Now let $\alpha \in \bar{Z}$. Since $\bar{Z} \subseteq \bar{C}$, we have $\tau_\alpha = \sup\ \{\tau_\alpha^\zeta | \zeta < \alpha \text{ and } \alpha \in Z_\zeta\}$. But if $\zeta < \zeta' < \alpha$, $\alpha \in Z_\zeta \cap Z_{\zeta'}$, then $\tau_\alpha^\zeta < f_{\zeta'}(\alpha) < \tau_\alpha^{\zeta'}$. Hence $\tau_\alpha = \sup\ \{f_\zeta(\alpha) | \zeta < \alpha \text{ and } \alpha \in Z_\zeta\}$. But each $f_\zeta(\alpha)$ has the form γ_i^α. Hence $\tau_\alpha = \gamma_\lambda^\alpha$ for some limit ordinal λ, since $\langle\gamma_i^\alpha | i \in On\rangle$ is a normal sequence. This shows that τ is as required.

Now let $\langle\tau^\zeta | \zeta < \kappa\rangle$ be a monotone sequence of $\tau \in E$ satisfying the conclusion of Lemma 3. Set $\tau = \sup_{\zeta<\kappa} \tau^\zeta$.

<u>Lemma 2.6</u>: Set $I = \{\tau^\zeta | \zeta < \kappa\}$ and let $A \in \mathfrak{P}(\tau) \cap K$. Then there is $\gamma \in \tau$ such that $I-\gamma \subseteq A$ or $I-\gamma \subseteq \tau-A$.

<u>Proof</u>: Let $\nu \in E$ such that $A \in K_\nu$. Argueing as in the proof of Lemma 2.5, we get $\bar{Z} \in U$ such that for $\alpha \in \bar{Z}$:

(a) $\tau_\alpha = \gamma_\lambda^\alpha$ for some limit λ, $\bar{\tau}, \tau, A \in \text{rug}\ \pi_\alpha^\nu$;

(b) $\{\zeta < \alpha | \exists i\ \tau_\alpha^\zeta = \gamma_i^\alpha\}$ is cofinal in α.

For $\alpha \in \bar{Z}$ set $N^\alpha = M_\lambda^\alpha$ and $B_\alpha = \{\gamma_i^\alpha | \alpha < \lambda\}$ where $\tau_{\alpha'} = \gamma_\lambda^\alpha$. Remember that the γ_i^α are the iteration points of the mouse M^α. Hence for every $X \in \mathfrak{P}(\gamma_\lambda^\alpha) \cap N^\alpha$ there is a $\delta < \gamma_\lambda^\alpha$ such that $B_\alpha - \delta \subseteq X$ or $B_\alpha - \delta \subseteq \gamma_\lambda^\alpha - X$. This applies in particular to $A_\alpha := (\pi_\alpha^\nu)^{-1}(A)$, since we have $A_\alpha \in N^\alpha$ by Lemma 2.4.

We may assume w.l.o.g. that A_α contains a final segment of B_α for every $\alpha \in \bar{Z}$. Using (b) we can define a regressive $g: \bar{Z} \to \kappa$ such that $B_\alpha - \tau_\alpha^{g(\alpha)} \subseteq A_\alpha$ for $\alpha \in Z$. But then g is bounded mod U by some $\gamma < \kappa$. Using the fact that $Y_\zeta = \{\alpha \in \bar{Z} | \tau_\alpha^\zeta \in B_\alpha\} \in U$, it follows immediately that $I - \gamma \subseteq A$.

We are now ready to finish Case 1. By Lemma 2.6 we can define an ultrafilter V over $\mathcal{P}(\tau) \cap K$ as follows:

$$A \in V \quad \text{iff} \quad \exists \gamma < \tau \quad I - \gamma \subseteq A.$$

V is ω-closed, since $\mathrm{cf}(\tau) = \kappa > \omega$. Hence the ultrapower of K by V is wellfounded.

<u>Case 2</u>: $W_\tau = \{\alpha \in C_\tau | K_{\tau_\alpha} = K_\alpha^\tau\} \in U$ for all $\tau \in E$.

<u>Case 2.1</u>: $(\kappa^+)^K < \kappa^+$.

Set $\kappa' = (\kappa^+)^K$. Fix $\tau \in E$. Clearly $\kappa' < \tau$, since $K_\tau \prec K_{\kappa^+}$. Let $\pi_\alpha^\tau(\alpha') = \kappa'$ for $\alpha \in W_\tau$. We first show that $\{\alpha \in W_\tau | \alpha' = (\alpha^+)^K\}$ belongs to U. Suppose not. For $\alpha \in W_\tau$ such that $\alpha' \neq (\alpha^+)^K$ pick $\delta_\alpha < \bar{\bar{\alpha}}^+$ such that $\bar{\bar{\alpha}}' \leq \alpha$ in K_{δ_α}. Pick $\bar{\tau} > \tau$, $\bar{\tau} \in E$, such that $Y = \{\alpha | \bar{\tau}_\alpha > \delta_\alpha\} \in U$. Choose $\alpha \in Y$ with $\tau \in \mathrm{rng}\, \pi_\alpha^{\bar{\tau}}$. Then α' is a cardinal in K_α^τ but not in $K_\alpha^{\bar{\tau}}$ which contradicts $K_\alpha^\tau \prec K_\alpha^{\bar{\tau}}$.

Set $Y = \{\alpha \in W_\tau | \alpha' = (\alpha^+)^K\}$. For $\alpha \in Y$ we can define a non-trivial ultrafilter over $\mathcal{P}(\alpha) \cap K$ by:

$$X \in V_\alpha \quad \text{iff} \quad \alpha \in \pi_\alpha^\tau(X).$$

It suffices to show that some V_α yields a wellfounded ultrapower of K.

Suppose not. Then for each $\alpha \in Y$ there is a sequence f_n^α such that $f_n^\alpha \in K$, $f_n^\alpha: \alpha \to K$ and $\{\nu < \alpha | f_{n+1}^\alpha(\nu) \in f_n^\alpha(\nu)\} \in V_\alpha$. But then we may take $f_n^\alpha \in K_{\bar{\bar{\alpha}}^+}$. To see this, let us assume $K \neq L$, since otherwise the argument is well-known. Then there is a mouse N such that $\{f_n^\alpha | n < \omega\} \subseteq N$.

Let $\tilde{N} \prec N$ such that $\{f_n^\alpha | n < \omega\} \cup \alpha \subseteq N$ and $\overline{\overline{\tilde{N}}} = \overline{\overline{\alpha}}$. Let $\pi: \bar{N} \tilde{\to} \tilde{N}$ where $\bar{N}$ is transitive. Then $\bar{N}$ is a mouse of cardinality $\overline{\overline{\alpha}}$, hence $\bar{N} \in K_{\overline{\overline{\alpha}}^+}$. But $\pi^{-1}(f_n^\alpha)$ is also a ϵ-descending sequence mod V_α.

Now choose $\delta_\alpha < \overline{\overline{\alpha}}^+$ such that $\{f_n^\alpha | n < \omega\} \subseteq K_{\delta_\alpha}$ and pick $\bar{\tau} > \tau$ such that $\{\alpha | \bar{\tau}_\alpha \geq \delta_\alpha\} \in U$. Then $Z = \{\alpha \in Y \cap W_{\bar{\tau}} | \tau \in \text{rng } \pi_\alpha^{\bar{\tau}}$ and $\bar{\tau}_\alpha \geq \delta_\alpha\} \in U$. Let $\alpha \in Z$ and set $X_n = \{\nu | f_{n+1}^\alpha(\nu) \in f_n^\alpha(\nu)\}$, $X_n^* = \pi_\alpha^{\bar{\tau}}(X_n)$ and $f_n^* = \pi_\alpha^{\bar{\tau}}(f_n^\alpha)$. Then $\alpha \in X_n^* = \{\nu | f_{n+1}^*(\nu) \in f_n^*(\nu)\}$, hence $f_{n+1}^*(\alpha) \in f_n^*(\alpha)$. Contradiction!

<u>Case 2.2</u>: $(\kappa^+)^K = \kappa^+$.

In this case we actually show that $U \cap L[U]$ is normal in $L[U]$. By a theorem of [3] it suffices to show that $\langle K_{\kappa^+}, U \cap K \rangle$ is amenable and $U \cap K$ is normal in K. This is equivalent to the following

<u>Claim</u>: There are arbitrarily large $\tau < \kappa^+$ such that $\langle K_\tau, U \cap K_\tau \rangle$ is amenable and $U \cap K_\tau$ is normal in $\langle K_\tau, U \cap K_\tau \rangle$.

Let $\gamma < \kappa^+$. We shall show that there is $\tau > \gamma$ satisfying the claim. For the moment, let $\tau \in E$ be arbitrary. Recall that
$W_\tau = \{\alpha \in C_\tau | K_\alpha^\tau = K_{\tau_\alpha}\} \in U$. For $\alpha \in W$ define:

$$V_\alpha^\tau = \{X \in \mathcal{P}(\alpha) \cap K_{\tau_\alpha} | \alpha \in \pi_\alpha^\tau(X)\}.$$

We first show

$$W_\tau = \{\alpha \in W_\tau | V_\alpha^\tau \in K\} \in U.$$

To see this pick $\bar{\tau} \in E$, $\bar{\tau} > \tau$. Then $Y = \{\alpha \in W_{\bar{\tau}} | \tau \in \text{rng } \pi_\alpha^{\bar{\tau}}\} \in U$. Let $\alpha \in Y$. Let $A = \langle A_\nu | \nu < \alpha \rangle \in K_{\bar{\tau}_\alpha}$ be an enumeration of $\mathcal{P}(\alpha) \cap K_{\tau_\alpha}$ and set $A^* = \pi_\alpha^{\bar{\tau}}(A)$. Then $V_\alpha^\tau = V_\alpha^{\bar{\tau}} \cap K_{\tau_\alpha} = \{A_\nu | \alpha \in A_\nu^*\} \in K$.
Note that we actually have $V_\alpha^\tau \in K_{\overline{\overline{\alpha}}^+}$ for $\alpha \in \tilde{W}_\tau$.

It is now obvious that we can apply the method of Lemma 2.5 to find $\tau \in E$ such that $\tau > \eta$, $cf(\tau) = \kappa$ and

$Y = \{\alpha \in W_\tau \mid \langle K_{\tau_\alpha}, V_\alpha^\tau \rangle$ is amenable$\} \in U$.

We shall show that τ satisfies our <u>Claim</u>.

By the definition of V_α^τ, we obviously have for $\alpha \in Y$:

V_α^τ is normal in $\langle K_{\tau_\alpha}, V_\alpha^\tau \rangle$.

For $\alpha \in Y$ choose $\delta_\alpha < \bar{\bar{\alpha}}^+$ such that $V_\alpha^\tau \in K_{\delta_\alpha}$. Pick $\bar{\tau} \in E$, $\bar{\tau} > \tau$, such that $\{\alpha \mid \delta_\alpha < \bar{\tau}_\alpha\} \in U$. Set $Z = \{\alpha \in Y \cap W_{\bar{\tau}} \mid \tau \in \text{rng}\, \pi_\alpha^\tau$ and $\bar{\tau}_\alpha > \delta_\alpha\}$. Then $Z \in U$. For $\alpha \in Z$ we have $V_\alpha^\tau \in K_\alpha^{\bar{\tau}}$, so set $V^\alpha = \pi_\alpha^{\bar{\tau}}(V_\alpha^\tau)$. Then:

V^α is normal on κ in $\langle K_\tau, V^\alpha \rangle$ and $\langle K_\tau, V^\alpha \rangle$ is amenable.

We need a final lemma

<u>Lemma 2.7</u>: Let $K_\tau \prec K_{\kappa^+}$ and $cf(\tau) > \omega$, $\tau < \kappa^+$. Then there is at most one V satisfying

V is normal on κ in $\langle K_\tau, V \rangle$ and $\langle K_\tau, V \rangle$ is amenable.

Let us first show how this finishes our proof. We now know that $V^\alpha = V^{\alpha'} = V$ for every $\alpha, \alpha' \in Z$. It follows immediately that every $X \in V$ contains a final segment of Z. But $Z \in U$. Hence $V = U \cap K_\tau$. So τ satisfies our <u>Claim</u>.

<u>Proof of Lemma 2.7</u>: Let V be normal on κ in $\langle K_\tau, V \rangle$ where $\langle K_\tau, V \rangle$ is amenable. It suffices to show:

<u>Claim</u>: There is a mouse $N = J_\beta^V$ at κ such that $\mathcal{P}(\kappa) \cap N = \mathcal{P}(\kappa) \cap K_\tau$.

We first show that the claim implies the lemma. Suppose not. Then we get distinct mice N, M at κ such that $\mathcal{P}(\kappa) \cap N = \mathcal{P}(\kappa) \cap M$. Let e.g. $N \lhd M$. Then $A_{N'} \in \mathcal{P}(\kappa) \cap M \subseteq N$. Contradiction!

We now prove the claim.

Set $M = J_\alpha^V = \bigcup\{J_{\nu+1}^V \mid J_\nu^V \in K_\tau\}$.

We closely follow an argument in the proof of Lemma 1.3 As there, we have $M \subseteq K_\tau$ but $M \notin K_\tau$. V is ω-closed, since $cf(\tau) > \omega$. Hence M is iterable. As in (7), (8) of the earlier proof we first get $\mathcal{P}(\kappa) \cap M \nsubseteq K_\zeta$ for $\zeta < \tau$ (using $K_\tau \prec K_{\kappa^+}$) and then $\mathcal{P}(\kappa) \cap M = \mathcal{P}(\kappa) \cap K_\tau$.

Since $\tau < \kappa^+$, we know that V is not normal in L[V]. Hence there is a least β such that $\mathcal{P}(\kappa) \cap \Sigma_\omega(J_\beta^V) \nsubseteq K_\tau$. Then $\beta \geq \alpha$ and $N = J_\beta^V$ is a mouse, since V is ω-closed. So N satisfies the claim.

This completes the proof of Theorem 2.1.

We strongly suspect, but cannot prove, that the assumption $\neg L^\mu$ precludes the existence of any weakly normal ultrafilter. A light modification of the foregoing argument does at least tell us, however, that assuming $\neg L^\mu$ there is no weakly normal ultrafilter on regular κ such that $2^{\underline{\kappa}} = \kappa$.

__Theorem 2.8__: Assume $\neg L^\mu$. Let n be regular such that $2^{\underline{\kappa}} = \kappa$. Then there is no weakly normal ultrafilter on κ.

We have to use a general property of weakly normal ultrafilters which was discovered by Ketonen (see [8]). For completeness, we provide a proof.

__Lemma 2.9__: Let U be weakly normal on κ, κ regular.

(a) Let $\langle u_\nu \mid \nu \in X\rangle$ be a regularity sequence for U.

Let $D \subseteq \kappa$ be unbounded in κ.

Then $\{\nu | \sup(D \cap u_\nu) = \nu\} \in U$.

(b) U has a regularity sequence $\langle u_\lambda | \lambda < \kappa, \lim(\lambda)\rangle$ such that $|u_\lambda| \leq cf(\lambda)$ for all λ.

Proof:

(a) Suppose not. Then the map $\nu \to \sup\,(D \cap u_\nu)$ is regressive mod U. Hence there is $\gamma < \kappa$ such that $Z \in U$, where $Z = \{\nu | D \cap u_\nu \subseteq \gamma\}$. Let $\delta \in D - \gamma$. Then $W = \{\gamma | \delta \in u_\gamma\} \in U$. But $Z \cap W = \emptyset$. Contradiction!

(b) Let $a_\lambda \subseteq \lambda$ such that $\sup a_\lambda = \lambda$ and $|a_\lambda| = \lambda$. Define $\eta_\zeta, \bar{\eta}_\zeta, f_\zeta$ by induction on $\zeta < \kappa$ as follows.

$$\bar{\eta}_\zeta = \sup_{\nu<\zeta} \eta_\zeta$$

$$f_\zeta(\lambda) = \min\,(a_\lambda - \bar{\eta}_\zeta) \quad \text{for } \lambda > \bar{\eta}_\zeta$$

$$\eta_\zeta = \min\,\{\eta | f_\zeta^{-1}\text{"}\,\eta \in U\}$$

Set $u_\lambda = \{\zeta < \lambda | \lambda \in dom(f) \text{ and } f_\zeta(\lambda) \in \eta_\zeta\}$. Then $\langle u_\lambda\rangle$ is a regularity sequence with the desired property.

Proof of Theorem 2.8: Let U be weakly normal on κ, where κ is regular and $2^{\underset{\sim}{\kappa}} = \kappa$. By Lemma 2.9 U has a regularity sequence $\langle u_\lambda | \lambda < \kappa, \lim(\lambda)\rangle$ such that $|u_\lambda| = cf(\lambda)$ and $\sup u_\lambda = \lambda$ for all λ. Let this sequence be fixed for the rest of the proof.

Define $E, f_\nu, A, Q_\tau, \tilde{Q}^\tau_\alpha, C_\tau$ exactly as in the proof of Theorem 2.1. We now require, however, that $H_\kappa = L_\kappa[A]$ and that $A \cap \kappa$ code $D \cap \kappa$. This is possible, since $2^{\underset{\sim}{\kappa}} = \kappa$. We also set $Q_o = \langle L_\kappa[A], A\cap\kappa\rangle$ and let C_o be the set of $\alpha < \kappa$ such that $\langle L_\alpha[A], A \cap \alpha\rangle \prec Q_o$. For $\alpha \in C_o$ we may set $\tilde{Q}^o_\alpha = \langle L_\alpha[A], A\cap\alpha\rangle$. For $\tau \in \{o\} \cup E$, $\alpha \in C_\tau$ we then define:

$$\hat{Q}^\tau_\alpha = \text{the smallest } Q \prec Q_\tau \text{ such that } u_\alpha \subseteq Q.$$

Then $\hat{Q}^\tau_\alpha \prec \tilde{Q}^\tau_\alpha \prec Q_\tau$ and $\alpha = \kappa \cap \tilde{Q}^\tau_\alpha = \sup\,(\kappa \cap \hat{Q}^\tau_\alpha)$. It is clear that, if $\alpha \in C_\tau$, then $\alpha \in C_o$ and $\hat{Q}^o_\alpha \subseteq \hat{Q}^\tau_\alpha$. Now set for $\tau \in \{o\} \cup E$:

$$C'_\tau = \{\alpha \in C_\tau | \hat{Q}^o_\alpha = \hat{Q}^\tau_\alpha \cap H_\kappa\}.$$

Lemma 2.10: $C_\tau^! \in U$.

Proof: Let $h: H_\kappa \to \kappa$ be a Q_o-definable bijection. Set $t_\beta = \{\langle i,x\rangle \mid x \in L_\beta[A]$ and $Q_\tau \models \varphi_i(x)\}$ for $\beta < \kappa$, where $\langle\varphi_i\rangle$ is a recursive enumeration of the formulas. Set $D = \{h(t_\beta) \mid \{h(t_\gamma) \mid \gamma < \beta\} \subseteq \beta\}$. Then D is unbounded in κ. So by Lemma 2.9 $X = \{\alpha \in C_\tau \mid \sup(D \cap u_\alpha) = \alpha\} \in U$. We shall show that $X \subseteq C_\tau^!$. Let $\alpha \in X$. Then $\alpha = \sup\{\beta < \alpha \mid h(t_\beta) \in u_\alpha\}$. Let $x \in H_\kappa \cap \hat{Q}_\alpha^\tau$. Then x is Q_τ-definable in a finite $v \in u_\alpha$. Choose $\beta < \alpha$ such that $v \subseteq \beta$ and $h(t_\beta) \in u_\alpha$. Then $t_\beta \in \hat{Q}_\alpha^o$ and there is $i < \omega$ such that

x = that x such that $\langle i,\langle x,v\rangle\rangle \in t_\beta$.

Hence $x \in \hat{Q}_\alpha^o$.

In the present proof, the structures $\hat{Q}_\alpha^\tau$ play the crucial role. So for $\tau \in \{0\} \cup E$, $\alpha \in C_\tau^!$, we now define Q_α^τ, $\hat{\pi}_\alpha^\tau$ by:

$\hat{\pi}_\alpha^\tau: Q_\alpha^\tau \cong \hat{Q}_\alpha^\tau$, where Q_α^τ is transitive,

and set:

$$K_\alpha^\tau = K^{Q_\alpha^\tau},\ \pi_\alpha^\tau = \hat{\pi}_\alpha^\tau \restriction K_\alpha^\tau,\ \pi_\alpha = \mathrm{On} \cap K_\alpha^\tau.$$

We finally set: $\bar{\alpha} = O_\alpha = \mathrm{On} \cap K_\alpha^o$.

Note that $\tau_\alpha < \mathrm{cf}(\alpha)^+$.

The following facts are easily verified:

(a) If $\alpha \in C_{\tau'}$, then $K_\alpha^o = K_{\bar{\alpha}}^{Q_\alpha^\tau}$ and $\pi_\alpha^o = \pi_\alpha^\tau \restriction K_\alpha^o$

(b) If $\alpha \in C_\tau^!$, $\eta \in E \cap \mathrm{rng}\,\pi_\alpha^\tau$,

then $K_\alpha^\eta = K_{\eta_\alpha}^{K_\alpha^\tau} \in K_\alpha^\tau$ and $\pi_\alpha^\eta = \pi_\alpha^\tau \restriction K_\alpha^\eta$

(c) Let $x \in K_\tau$. Then $\{\alpha \mid x \in \mathrm{rng}\,\pi_\alpha^\tau\} \in U$.

(For (c) assume w.l.o.g. $x \in K_\tau \cap \mathrm{On}$. Let $x = f_\tau(\delta)$. Then $x \in \mathrm{rng}\,\pi_\alpha^\tau$ if $\delta \in u_\alpha$.)

We have the following substitute for Lemma 2.3

Lemma 2.11: Let $W \in U$, $f \in \prod_{\alpha \in W} cf(\alpha)^+$. Then there is $\tau \in E$ such that $\{\alpha | \tau_\alpha \geq f(\alpha)\} \in U$.

Proof: Suppose not. For $\alpha \in W$ let h_α inject $f(\alpha)$ into $cf(\alpha)$. For $\tau \in E$ set $Y_\tau = \{\alpha \in W \cap C'_\tau | \tau_\alpha < f(\alpha)\}$ and define $g_\tau \in \prod_{\alpha \in Y_\tau}$ by $g_\tau(\alpha) = h_\alpha(\tau_\alpha)$. By (b)(c) above, the g_τ are pairwise distinct mod U. Hence some $g_{\tau*}$ has κ many predecessors (mod U) g_{τ^ζ} $(\zeta < \kappa)$. Choose $\bar{\tau} \in E$ such that $\bar{\tau} > \tau^*, \tau^\zeta$ and set for $\alpha \in Y_{\tau^*} \cap C'_{\bar{\tau}} =: Y$:

$$v_\alpha = \{\zeta < \alpha | \tau^*, \tau^\zeta \in \operatorname{rng} \pi^{\bar{\tau}}_\alpha \text{ and } g_{\tau^\zeta}(\alpha) < g_{\tau^*}(\alpha)\} .$$

Then $\langle v_\alpha | \alpha \in Y\rangle$ is a regularity sequence for U. But $\bar{\bar{v}}_\alpha \leq g_{\tau^*}(\alpha) < cf(\alpha)$, hence $\sup v_\alpha < \alpha$ for all $\alpha \in Y$. This contradicts Lemma 2.9.

We now prove Theorem 2.8 by cases, closely imitating the proof of Theorem 2.1.

Case 1: There is $\tau \in E$ such that $\{\alpha \in C'_\tau | K^\tau_\alpha \neq K_{\tau_\alpha}\} \in U$.
Repeat the argument of Case 1 in Theorem 2.1. The repetition is almost literal, but uses the sets C'_τ and the points $cf(\alpha)^+$ in place of C_τ, $\bar{\bar{\alpha}}^+$.

Case 2: Case 1 fails.

Case 2.1: $\{\alpha \in C_0 | \pi^0_\alpha \neq \mathrm{id} \restriction K_\alpha\} \in U$.
Let $W = \{\alpha \in C_0 | \pi^0_\alpha \neq \mathrm{id} \restriction K_\alpha\}$. For $\alpha \in W$ let δ_α be the first point moved by π^0_α. Set $Y = \{\alpha \in W | \bar{\alpha}$ is a cardinal in $K\}$. We first show that $Y \in U$. Suppose not. For $\alpha \in W - Y$ there is $\gamma_\alpha < cf(\alpha)^+$ such that $\bar{\alpha}$ is not a cardinal in K_{γ_α}. Now pick τ such that $\{\alpha | \tau_\alpha \geq \gamma_\alpha\} \in U$. Let $Z = \{\alpha \in (W-Y) \cap C'_\tau | K_{\tau_\alpha} = K^\tau_\alpha$ and $\tau_\alpha \geq \gamma_\alpha\}$. Then $Z \in U$. Let $\alpha \in Z$. Then $\bar{\alpha}$ is regular in K^τ_α, since $\pi^\tau_\alpha : K^\tau_\alpha \longrightarrow_{\Sigma_\omega} K_{\kappa+}$ and $\pi^\tau_\alpha(\bar{\alpha}) = \kappa$. But $K^\tau_\alpha = K_{\tau_\alpha} \supseteq K_{\gamma_\alpha}$. Hence $\bar{\alpha}$ is not a cardinal in K^τ_α, which is a contradiction.

This proves $Y \in U$. But for $\alpha \in Y$ we clearly have $(\delta_\alpha^+)^K \leq \bar{\alpha}$. Hence we may define a non-trivial ultrafilter over $\mathcal{P}(\delta_\alpha) \cap K$ by:

$X \in V_\alpha$ iff $\delta_\alpha \in \pi_\alpha^o(X)$.

An imitation of Case 2.1, Theorem 2.1 shows that some V_α yields a well founded ultrapower of K.

<u>Case 2.2</u>: Case 2.1 fails.

In this case we have $\{\alpha < \kappa \mid \alpha \text{ regular}\} \in U$. Hence U is not (γ,κ)-regular for any $\gamma < \kappa$. So the conclusion follows from Theorem 2.1.

§ 3. Σ^1_3 - Absoluteness

Shoenfield showed that Σ^1_2 statements are absolute in L. His theorem does not extend to Σ^1_3 statements, since $\exists a\ a \notin L$ is itself Σ^1_3. On the other hand, the statement $\exists a\ a \notin K$ is Σ^1_4 rather than Σ^1_3. It is tempting, therefore, to conjecture that K is Σ^1_3-absolute. Clearly this cannot be the case if K is too small - e.g. if K = L. Nor can it be the case if V is too much larger than K - e.g. if $0^\dagger$ exists, since $0^\dagger$ is the unique solution of a Π^1_2 condition. Within these limits, however, we obtain a modest extension of Shoenfield's theorem.

Theorem 3.1: Let A be Π^1_2. Assume A(a), where $a^\#$ exists but $L[a] \models \neg L^\mu$. Then $\exists a \in K\ A(a)$.

We immediately get:

Corollary 3.2: Assume $\neg 0^\dagger$. Let A be Π^1_2. Then
$\exists a (A(a) \wedge a^\#$ exists$) \rightarrow \exists a \in K\ A(a)$.

This in turn yields:

Corollary 3.3: Assume $\neg 0^\dagger$ but that the reals are closed under $\#$. Then K is Σ^1_3-absolute.

As a corollary of the proof of Theorem 3.1, we shall obtain:

Theorem 3.4: Let A be Π^1_2. Let M be a mouse and assume A(a), where $M \notin L[a]$. Then $\exists a \in L[M]\ A(a)$.

Harrington and Kechris proved this for the case $M = 0^\#$ (see [5]).

The proof of Theorem 3.1 stretches over several sublemmas. From now on assume A(a), $L[a] \models \neg L^{\mu}$, $a^{\#}$ exists. Set $K^+ = K$ if $\neg L^{\mu}$. Otherwise set $K^+ = L[V_o]$ where V_o is normal on κ_o and κ_o is the least ordinal which is measurable in an inner model.

Lemma 3.5: There is an iterable premouse M such that:

(a) $M \in K^+$

(b) Let M_i be the iterates of M with iteration points κ_i. Then $H_{\kappa_i}^{M_i} = K_{\kappa_i}^{L[a]} \in M_i$.

Proof:

Case 1: $K^{L[a]} = K$.

Since $a^{\#}$ exists, there is a nontrivial embedding of L[a] into itself, hence of K into itself. Hence $K^+ = L[V_o]$ and we can take $M = J_{\kappa_o^+}^{V_o}$ setting $\kappa_o^+ = (\kappa_o^+)^K$.

Case 2: $K^{L[a]} \neq K$.

Then we can apply Lemma 1.4 and take the $\triangleleft$-least core mouse M such that $M \notin L[a]$. (Note that the $H_{\kappa_i}^{M_i}$ are the same whether we regard the M_i as the iterates or the mouse iterates of M).

From now on let $M = J_{\alpha}^{V}$ be fixed with V normal on κ in M. Let $M_i = J_{\alpha_i}^{V_i}$ be the iterates with iteration points κ_i and iteration maps π_{ij}. Set $C = \{\kappa_i \mid i \in On\}$. Finally set $K' = K^{L[a]}$. Then we have:

(1) $H_{\kappa_i}^{M_i} = K'_{\kappa_i}$

(2) $\mathcal{P}(K'_{\kappa_i}) \cap K' = \mathcal{P}(K'_{\kappa_i}) \cap M_i$

(since $\mathcal{P}(K'_{\kappa_i}) \cap M_i = \mathcal{P}(K'_{\kappa_i}) \cap M_{i+1}$)

(3) κ_i is inaccessible in K'.

Not let I be the canonical indiscernibles for L[a] and $\langle \tau_i | i < \omega \rangle$ the monotone enumeration. Let $b_{ij}: L[a] \longrightarrow_{\Sigma_1} L[a]$ be defined by:

$b_{ij} \upharpoonright \tau_i = \mathrm{id} \upharpoonright \tau_i$, $b_{ij}(\tau_{i+a}) = \tau_{j+h}$.

Define an ultrafilter U_i on $\mathcal{P}(\tau_i) \cap L[a]$ by:

$$X \in U_i \quad \text{iff} \quad \tau_i \in b_{ij}(X) \quad (i < j).$$

U_i is obviously normal in $\langle L[a], U_i \rangle$. It is known that $\langle L_{\tau_i^+}[a], U_i \rangle$ is amenable (setting $\tau_i^+ = (\tau_i^+)^{L[a]}$) and that $\langle L[a], U_i \rangle$ is the i-th iterate of $\langle L[a], U_o \rangle$, the b_{ij} being the iteration maps. We use the ultrafilters U_i to prove.

<u>Lemma 3.6</u>: $(\tau_i^+)^{K'} = (\tau_i^+)^{L[a]}$.

<u>Proof</u>: Set $\tau_i^+ = (\tau_i^+)^{L[a]}$, $\tau_i' = (\tau_i^+)^{K'}$. Suppose $\tau_i' < \tau_i^+$. Since $\langle L_{\tau_i^+}[a], U_i \rangle$ is amenable we then have $U_i' \in L[a]$, where $U_i' = U_i \cap K'$. A standard argument shows, however, that $\langle K'_{\tau_i'}, U_i' \rangle$ is amenable. Moreover, $\mathrm{cf}(\tau_i') = \tau_i$ in L[a], since τ_i is inaccessible in L[a] and the covering lemma holds in L[a] with respect to K' (see [3]). But then U_i' is ω-complete in L[a]. Hence, in L[a], we have by another theorem of [3] that U_i' is normal in $L[U_i']$. This contradicts our assumption that $L[a] \models \neg L^{\mu}$.

Now set $W = \mathcal{P}(K_\kappa') \cap K'$. For $X \in W$ set $\tilde{X} = \bigcup_i \pi_{oi}(X)$. Clearly $K' = \bigcup_{x \in W} \tilde{X}$.

<u>Lemma 3.7</u>: There is i_o such that if $i_o \le i \le j$, then

$$b_{ij}(\tilde{X} \cap V_\nu) = \tilde{X} \cap V_{b_{ij}(\nu)} \quad \text{for } X \in W,\ \nu \in \mathrm{On}.$$

<u>Proof</u>: Set $\Gamma = \{\eta | \eta = \tau_\eta \text{ and } \mathrm{cf}(\eta) > \kappa\}$. Clearly, Γ is a stationary class. Let $\eta \in \Gamma$. By a Foder type argument for each $x \in L[a]$ there is

$\zeta < \eta$ such that $b_{ij}(x) = x$ for $\zeta \leq i \leq j \leq \eta$. Since $\bar{\bar{W}} \leq \kappa < \mathrm{cf}(\eta)$, there is $\zeta_\eta < \eta$ such that $\zeta_\eta \leq i \leq j < \eta$ implies $b_{ij}(X \cap V_\eta) = X \cap V_\eta$ for all $X \in W$. The map $\eta \to \zeta_\eta$ is regressive on Γ, hence constant on a stationary $\Gamma' \subset \Gamma$. Let $\zeta_\eta = i_o$ for $\eta \in \Gamma'$. We show that i_o is as required. Let $i_o \leq i \leq j$, $X \in W$, $\nu \in \mathrm{On}$. Pick $\eta \in \Gamma'$ such that $j, \nu < \eta$. Then:
$b_{ij}(\tilde{X} \cap V_\nu) = b_{ij}((\tilde{X} \cap V_\eta) \cap V_\nu) = (\tilde{X} \cap V_\eta) \cap V_{b_{ij}(\nu)} = \tilde{X} \cap V_{b_{ij}(\nu)}$.

<u>Corollary 3.8</u>: Let $i_o \leq i \leq j$. Then

(a) $\eta \in C$ iff $b_{ij}(\eta) \in C$

(b) $\tau_i \in C$.

<u>Proof</u>: (a) is immediate by the Lemma and $C = \bigcap_{X \in V} \tilde{X}$.
To prove (b) pick $\eta = \tau_\eta = \kappa_\eta$ such that $\eta \geq i$. Then $b_{i\eta}(\tau_i) = \eta \in C$, hence $\tau_i \in C$ by (a).

Every ordinal has the form $\tilde{f}(\vec{\gamma})$ for some $\vec{\gamma} \in C$, $f \in W$ such that $f: \kappa^n \to \kappa$. By Lemma 3.7 $b_{ij}(\tilde{f}(\vec{\gamma})) = \tilde{f}(b_{ij}(\vec{\gamma}))$ for $i_o \leq i \leq j$. Thus the action of b_{ij} upon the ordinals is determined by its action on C. We now prove a version of Paris' "patterns of indiscernibles" lemma (see [5]) to show that this action is very simple indeed.

<u>Lemma 3.9</u>: Let $\tau_{i_o} = \kappa_\alpha$. There is β such that

(a) $\tau_{i_o+j} = \kappa_{\alpha+\beta j}$

(b) $b_{i_o+i, i_o+j}(\kappa_{\alpha+\beta i+\nu}) = \kappa_{\alpha+\beta j+\nu}$.

<u>Proof</u>: Set $C_i = C \cap [\tau_i, \tau_{i+1})$. It suffices to show that for $i_o \leq i \leq j$

(*) $b_{ij}''C_{i+h} = C_{j+h}$.

But this follows from

(**) $b_{ij}''C_i = C_j$

since then $b_{ij}''C_{i+h} = b_{ij} b_{i,j+h}''C_i = b_{i,j+h}''C_i = C_{j+h}$.

(**) in turn follows from

(***) $b_{i,i+1}"C_i = C_{i+1}$.

To see this, assume (***). We prove (**) by induction on $j-i$. The case $j = i$ is trivial.

For $j = k+1$, $k \geq i$ we have

$b_{i,k+1}"C_i = b_{k,k+1}\, b_{ik}"C_i = b''_{k,k+1}C_k = C_{k+1}$.

For $j = \lambda$ such that $\lim(\lambda)$ it is enough to note that

C_λ, $\langle b_{i\lambda} \restriction C_i \mid i_o \leq i < \lambda\rangle$ is the direct limit of $\langle C_i \mid i_o \leq i < \lambda\rangle$, $\langle b_{ih} \restriction C_i \mid i_o \leq i \leq h < \lambda\rangle$.

We now prove (***). We recall that $\langle L[a], U_{i+1}\rangle$ is the ultrapower of $\langle L[a], U_i\rangle$. Hence each element of $L[a]$ has the form $b_{i,i+1}(f)(\tau_i)$ for an $f\colon \tau_i \to L[a]$, $f \in L[a]$ and

$$L[a] \models \varphi(b_{i,i+1}(\vec{f})(\tau_i)) \longleftrightarrow \{\nu < \tau_i \mid L[a] \models \varphi(\vec{f}(\nu))\} \in U_i.$$

We argue by contradiction. So let $\gamma \in C_{i+1} - \operatorname{rng}(b_{i,i+1})$. Let $\gamma = b_{i,i+1}(f)(\tau_i)$, where $f\colon \tau_i \to \tau_{i+1}$. Since $\gamma \notin \operatorname{rng}(b_{i,i+1})$, f is not constant on any $X \in U_i$. But then f is monotone on an $X \in U_i$ and we may take it to be monotone everywhere. Set $f_n = b_{i,i+n}(f)$ for $n \leq \omega$. Since $f_1(\tau_i) \in C \cap \tau_{i+2} = \bigcap_{X \in V} \tilde{X} \cap \tau_{i+2}$, we have:

(1) $\{\nu < \tau_i \mid f(\nu) \in \tilde{X} \cap \tau_i\} \in U_i$ for $X \in V$.

But then we get applying $b_{i,i+n}$:

(2) $\{\nu < \tau_{i+n} \mid f_n(\nu) \in \tilde{X} \cap \tau_{i+n}\} \in U_{i+n}$ $(n < \omega, X \in V)$.

This in turn implies

(3) $f_{n+1}(\tau_{i+n}) \in \bigcap_{X \in V} \tilde{X} \cap \tau_{i+n+1} \subsetneq C$ $(n < \omega)$.

Set $\gamma_n = f_{n+1}(\tau_{i+n})$, $\delta_n = b_{i+n+1,i+\omega}(\gamma_n) = f_\omega(\tau_{i+n})$. Finally set $\delta = \sup_{n<\omega} \delta_n$. Then $\delta_n \in C$; hence $\delta \in C$ since it is a limit point of C. But $\delta = \sup f''_\omega\, \tau_{i+\omega}$, since f_ω is monotone and $\sup_{n<\omega} \tau_{i+n} = \tau_{i+\omega}$. Hence

$cf(\delta) = \tau_{i+\omega}$ in $L[a]$. But δ is inaccessible in K' and $\bar{\bar{\delta}} > \tau_{i+\omega}$ in $L[a]$, since $(\tau^+_{i+\omega})^{L[a]} = (\tau^+_{i+\omega})^{K'}$ by Lemma 3.6. Hence the covering lemma for K' fails in $L[a]$. Contradiction!

Now define an equivalence relation $\sim\ = \ \sim_{M,\alpha,\beta}$ on the set of increasing tupels from C as follows:

$<\kappa_{i_0}, \ldots, \kappa_{i_{n-1}}> \sim <\kappa_{j_0},\ldots,j_{m-1}>$ iff $n = m$ and

(a) $i_k < \alpha \rightarrow i_k = j_k$

(b) $\alpha \le i_k \rightarrow (\alpha \le j_k$ and $\left(\frac{i_k-\alpha}{\beta}\right) = \left(\frac{j_k-\alpha}{\beta}\right)$)

(c) $\alpha \le i_k \le i_h$ and $\left[\frac{i_k-\alpha}{\beta}\right] = \left[\frac{i_h-\alpha}{\beta}\right] \rightarrow \left[\frac{j_k-\alpha}{\beta}\right] = \left[\frac{j_h-\alpha}{\beta}\right]$

(where $\gamma = \beta \cdot \left[\frac{\gamma}{\beta}\right] + \left(\frac{\gamma}{\beta}\right)$, $\left(\frac{\gamma}{\beta}\right) < \beta$).

By standard methods we then get:

<u>Corollary 3.10</u>: Let $<\vec{\gamma}>, <\vec{\delta}> \in [C]^n$ such that $<\vec{\gamma}> \sim <\vec{\delta}>$. Let $f_i \in W$, $f_i: \kappa^n \rightarrow \kappa$ for $i = 1,\ldots,m$. Then

$$L[a] \models \varphi(\vec{\vec{f}}(\vec{\gamma})) \longleftrightarrow L[a] \models \varphi(\vec{\vec{f}}(\vec{\delta})).$$

Now let M be an arbitrary iterable premouse at some κ and α,β arbitrary ordinals. Set $H = H_{M,\alpha,\beta} = L_\delta[M]$, where δ is the least $\delta > \alpha,\beta,On \cap M$ such that $L_\delta[M]$ is admissible. Let $<M_i,\pi_{ij},\kappa_i>$ be the iteration of M and set $C = \{\kappa_i \mid i < \infty\}$. Define $\sim\ = \ \sim_{M,\alpha,\beta}$ on $[C]^{<\omega}$ as before. Let $\bar{\kappa} = \bar{\kappa}_{M,\alpha,\beta} = \kappa_{\alpha+\beta\omega}$. Then $\bar{\kappa}, C \cap \bar{\kappa} \in H$; hence $(\sim\restriction[C \cap \bar{\kappa}]^{<\omega}) \in H$. Set $W = W_M$ = the set of $f \in M$ such that $f: \kappa^n \rightarrow \kappa$ for some n. For $f \in W$ set $\tilde{f} = \bigcup_i \pi_{oi}(f)$ and $\tilde{f}_\nu = \tilde{f}\restriction\nu^n$. Then $\tilde{f}_{\bar{\kappa}} = \pi_{o,\alpha+\beta\omega}(f)$ and $<\tilde{f}_{\bar{\kappa}} \mid f \in W> \in H$.

Now define an infinitary H-language $\mathcal{L} = \mathcal{L}_{M,\alpha,\beta}$ as follows.

The only predicate symbols of $\mathcal{L}$ are $=,\in$. $\mathcal{L}$ has constant names $\underline{x}$ $(x \in H)$ and an additional constant $\dot{a}$. As axioms we take ZF^- together with

(a) $\nu \in \underline{x} \leftrightarrow \bigvee\!\!\!\bigvee_{z \in X} \nu = \underline{z}$

(b) $\mathring{a} \subseteq \omega$

(c) $L_{\bar{\kappa}}[\mathring{a}] \models A(\mathring{a})$

(d) If $\langle\vec{\gamma}\rangle, \langle\vec{\delta}\rangle \in [C \cap \bar{\kappa}]^{<\omega}$, $\langle\vec{\gamma}\rangle \sim \langle\vec{\delta}\rangle$ and $f_i \in W$ such that $f_i: \kappa^n \to \kappa$ $(i = 1,\ldots,m)$, then $L_{\bar{\kappa}}[\mathring{a}] \models \varphi(\vec{\tilde{f}}_{\bar{\kappa}}(\vec{\gamma})) \leftrightarrow L_{\bar{\kappa}}[\mathring{a}] \models \varphi(\vec{\tilde{f}}_{\bar{\kappa}}(\vec{\delta}))$ for every formula φ.

An immediate consequence of the foregoing lemmas is

Lemma 3.11: There is $\langle M,\alpha,\beta\rangle \in K^+$ such that $\mathcal{L}_{M,\alpha,\beta}$ is consistent.

But this gives

Corollary 3.12: There is $\langle M,\alpha,\beta\rangle \in K$ such that M,α,β are countable in K and $\mathcal{L}_{M,\alpha,\beta}$ is consistent.

Proof: Let $\langle M,\alpha,\beta\rangle$ be as in the lemma. Let $X \prec H$, $\bar{\bar{X}} = \omega$ in K^+ such that $M,\alpha,\beta \in X$. Let $b: \bar{H} \xleftrightarrow{\sim} X$ where $\bar{H}$ is transitive. Set $\bar{M} = b^{-1}(M)$, $\bar{\alpha} = b^{-1}(\alpha)$, $\bar{\beta} = b^{-1}(\beta)$. Then $\bar{H} = H_{\bar{M},\bar{\alpha},\bar{\beta}}$ and $\mathcal{L}_{\bar{M},\bar{\alpha},\bar{\beta}}$ is consistent. But $\bar{M},\bar{\alpha},\bar{\beta}$ are countable in K^+. We claim that $\langle\bar{M},\bar{\alpha},\bar{\beta}\rangle \in K$. If $K = K^+$, there is nothing to be proved. Otherwise $K^+ = L[V_0]$ and $\langle\bar{M},\bar{\alpha},\bar{\beta}\rangle \in H_{\kappa_0}^{L[V_0]} = K_{\kappa_0}$. The same argument shows that $\bar{M},\bar{\alpha},\bar{\beta}$ are countable in K.

We are now ready to finish the proof of Theorem 3.1. Let M,α,β and $\mathcal{L} = \mathcal{L}_{M,\alpha,\beta}$ be as in Corollary 3.12. By the Barwise compactness theorem let $\mathfrak{A} \in K$ be a model of $\mathcal{L}$. We may assume that the well founded part of $\mathfrak{A}$ is transitive. Let $\bar{a}$ be the $\mathfrak{A}$-interpretation of $\mathring{a}$. Then $L_{\bar{\kappa}}[\bar{a}] \models A(\bar{a})$. It suffices to show:

<u>Claim</u>: $L[\bar{a}] \models A(\bar{a})$

Let $\mathcal{L}'$ be the ZF-language with the constant $\mathring{a}$ and ordinal constants $\underline{\nu}$ $(\nu \in On)$. Define a class S of $\mathcal{L}'$-sentences as follows:

Let $\langle\gamma\rangle \in [C]^n$, $f_i \in W$, $f_i: \kappa^n \to \kappa$ $(i=1,\ldots,m)$.

Then:

$$\ulcorner\varphi(\underline{\vec{f}(\vec{\gamma})})\urcorner \in S$$

$$\text{iff} \quad \exists\langle\vec{\delta}\rangle \in [C \cap \bar{\kappa}]^n\left(\langle\vec{\delta}\rangle \sim \langle\gamma\rangle \text{ and } L_{\bar{\kappa}}[\bar{a}] \models \varphi(\vec{\tilde{f}}(\vec{\delta}))\right)$$

Indiscernibility arguments show that this is a correct definition and that

(1) S is a consistent, deductively closed class of sentences

(2) $\ulcorner\varphi(\underline{\nu})\urcorner \in S$ iff $L_{\bar{\kappa}}[\bar{a}] \models \varphi(\vec{\nu})$ for $\vec{\nu} < \bar{\kappa}$

(3) $\ulcorner\exists x\, \varphi(x)\urcorner \in S$ iff $\exists t \in T$ $\ulcorner\varphi(t)\urcorner \in S$

where T is the class of ι-terms

(4) $\ulcorner\exists x \in On\, \varphi(x)\urcorner \in S$ iff $\exists\nu$ $\ulcorner\varphi(\underline{\nu})\urcorner \in S$

Now let $\mathfrak{L}$ be the term model of S. By (4), $\in_{\mathfrak{L}}$ is well founded and the rank of [t] in $\in_{\mathfrak{L}}$ is ν where $\ulcorner rn(t) = \underline{\nu}\urcorner \in S$ ([t] being the equivalence set of a term t). Hence $\mathfrak{L}$ is isomorphic to a transitive model Q. But then $\bar{a}$ is the Q-interpretation of $\mathring{a}$ and $L_{\bar{\kappa}}[\bar{a}] \prec Q$ by (2). Hence $Q = L[\bar{a}]$ and $L[\bar{a}] \models A(\bar{a})$. This finishes the proof of Theorem 3.1.

In conclusion we prove Theorem 3.4, mentioned at the outset. Let a,M be as in the hypothesis of that theorem. Assume w.l.o.g. that M is the $\triangleleft$-least core mouse. $M \notin L[a]$. We first show that a exists. Suppose not. Then the covering lemma holds for L[a]. But $L[a] \models \neg L^{\mu}$, since $K^{L[a]} \neq K$; hence in L[a] the covering lemma holds with respect to $K' = K^{L[a]}$. Hence the covering lemma holds with respect to K'. This is nonsense, since the mouse M enables us to construct a nontrivial

Σ_1-embedding of K' into itself. Thus $a^{\#}$ exists. M is then the iterable premouse used in the above proof which shows $\exists a \in L[M]\ A(a)$.

§ 4 Decomposability of Ultrafilters

At first we repeat some definitions and elementary results for ultrafilters. Let U be an ultrafilter on some cardinal κ. Let δ be a cardinal.

U is called δ-descendingly incomplete iff there is a sequence $\langle A_\nu \mid \nu < \delta \rangle \in U^\delta$ such that $\nu < \mu < \delta$ implies $A_\nu \supsetneq A_\mu$ and $\bigcap_{\nu<\delta} A_\nu = \emptyset$.

U is called δ-decomposable iff there is a partition $\langle a_\nu \mid \nu < \delta \rangle$ such that $\bigcup_{\nu<\delta} a_\nu \in U$ and for all $S \subseteq \delta$ with $\overline{\overline{S}} < \delta$ $\bigcup_{\gamma \in S} a_\nu \notin U$. U is δ-descendingly incomplete iff U is cf δ-decomposable. For more on these notions see [9] and [12].

U is called uniform iff for all $A, B \in U$ $\overline{\overline{A}} = \overline{\overline{B}}$.

An ultrafilter U' is below U in the Rudin-Keisler-ordering U' $\leq_{RK}$ U iff there is a function $f\colon \kappa \to \bigcup U'$ such that $U' = \{f''A \mid A \in U\}$. U is δ-decomposable iff there is a uniform ultrafilter U' on δ below U in the Rudin-Keisler-ordering.

Let λ be a cardinal. Slightly extending the notion of § 2 we call a sequence $\langle u_\nu \mid \nu \in X \rangle$ a λ-regularity sequence for U iff $X \subseteq \kappa$, $u_\nu \subseteq \nu$ for $\nu \in X$ and $\{\nu \mid \eta \in u_\nu\} \in U$ for all $\eta < \lambda$. Hence the κ-regularity sequences for U are just the regularity sequences for U in the sense of § 2.

Let γ be a cardinal. U is called (γ,λ)-regular iff there is a subset $H \subseteq U$ such that $\overline{\overline{H}} = \lambda$ and for all $H' \subseteq H$ $\overline{\overline{H'}} \geq \gamma$ implies $\bigcap H' = \emptyset$.

U is (γ,λ)-regular iff there is a λ-regularity sequence $\langle u_\nu \mid \nu \in X \rangle$

for U such that $\overline{\overline{u_\nu}} < \gamma$ for all $\nu \in X$. Hence for $\lambda = \kappa$ the notion of (λ,κ)-regularity of § 2 coincides with the notion introduced here. Let γ', λ' be cardinals such that $\gamma \leq \gamma' \leq \lambda' \leq \lambda$. Then the (γ,λ)-regularity of U implies the (γ',λ')-regularity of U. The (ω,λ)-regularity of U is just what is called λ-regularity in model theorey, see e.g. [2]. U is λ-decomposable iff it is (λ,λ)-regular for regular cardinals λ. If U is (γ,λ)-regular, then it is δ-decomposable for all δ such that $\gamma \leq cf(\delta) \leq \delta \leq \lambda$. Let $U' \leq_{RK} U$ be δ-deocmposable ((γ,λ)-regular). Then U is δ-decomposable ((γ,λ)-regular).

Now let U be a uniform ultrafilter on κ. $f \in \kappa^\kappa$ is a first function of U iff for all $g \in \kappa^\kappa$ $g < f$ (mod U) implies $g''y \subsetneq \tau$ for some $y \in U$ and $\tau < \kappa$. Then U is weakly normal iff $id \restriction \kappa$ is a first function of U. (See § 2). If $g \in \kappa^\kappa$ is a first function of U, then $U' = \{g''A \mid A \in U\}$ is a weakly normal ultrafilter $\leq_{RK} U$ on κ.

The following lemma is a theorem of Kanamori and is proved in [6].

<u>Lemma 4.1</u>: Let U be a uniform ultrafilter on κ without first function for U. Then U is (ω,λ)-regular for all cardinals $\lambda < \kappa$.

The next lemma is a result of Kunen and Prikry and follows easily from the theorems of [9] and the preceeding remarks.

<u>Lemma 4.2</u>: Let δ be a cardinal and U an ultrafilter which is δ^+-decomposable. Then U is $f(\delta)$-decomposable.

The following result was proved in 1974 under the stronger assumptions that O does not exist ($\neg 0^\#$) and all limit cardinals are strong limit cardinals (L C H). In 1976 it was proved without assuming L C H applying results analogous to those of § 2 with $\neg O^{\#}$ instead of $\neg L^{\mu}$. This proof now gives with the stronger results of § 2 the following result on decomposability.

Theorem 4.3: Assume $\neg L^{\mu}$. Let κ be a regular cardinal. Let U be a uniform ultrafilter on κ. Then U is δ-decomposable for all regular cardinals $\delta \leq \kappa$.

Proof: Evidently U is not δ-complete for any cardinal $\delta > \omega$ and hence ω-decomposable. Now we proceed by induction on κ.

Case 1: Let κ be regular and let U be uniform on κ^{+}. Then U is κ^{+}-decomposable and by Lemma 4.2 and the preceeding remarks there is some uniform ultrafilter $U' \leq_{RK} U$ on κ. Then by the induction hypothesis U' is δ-decomposable for all $\delta \leq \kappa$. But then U is δ-decomposable for all $\delta \leq \kappa^{+}$ by the remarks in the beginning of § 4.

Case 2: Let κ be a regular limit cardinal and let U be a uniform ultrafilter on κ. Assume that U has no first function. Then by Lemma 4.2 U is (ω,λ)-regular for all cardinals $\lambda < \kappa$. But then by the preceeding remarks U is λ-decomposable for all $\lambda < \kappa$. Now assume that U has a first function. But then there is a weakly normal ultrafilter $U' \leq_{RK} U$ on κ, and by theorem 2.1 U' is (γ,κ)-regular for some $\gamma < \kappa$. Now let $\delta < \kappa$ be a cardinal. Then there is a regular cardinal $\delta \leq \delta' < \kappa$ such that $\gamma \leq \delta' \leq \kappa$ since κ is a regular limit cardinal. But now the remarks show that there is a uniform ultrafilter $U_{\delta'} \leq_{RK} U'$ on δ', which by induction hypothesis is δ-decomposable. But then $U_{\delta} \leq_{RK} U' \leq_{RK} U$ and U is δ-decomposable.

Remark: Clearly a modification of this argument also would cover case 1.

Case 3: Set $\kappa = \beta^+$ for some singular cardinal β. Here the proof makes extensive use of the methods of [8] and [12]. We need the following lemma:

Lemma 4.4: Let β be a singular cardinal and U a weakly normal ultrafilter on β^+. Assume $\neg L^\mu$. Then one of the following conditions holds:

(i) U is (ω_1, β^+)-regular

(ii) $\bigcup\{\nu | U$ is ν-decomposable $\wedge$ ν regular$\} = \beta$.

Clearly this lemma settles case 3. Let U be a uniform ultrafilter on β^+. If U does not have a first function, then by lemma 4.1, U is (ω, β)-regular and so δ-decomposable for all $\delta \leq \beta^+$. If U has a first function, then there is a weakly normal $V \leq_{RK} U$ on β^+. Now (i) of lemma 4.4 implies that V and hence U are δ-decomposable for all $\delta \leq \beta^+$ such that $\omega_1 \subseteq cf(\delta)$ and $\delta = \omega$. But (ii) and the induction hypothesis show that V and U are δ-decomposable for all regular $\delta \leq \beta^+$.

The proof of lemma 4.4 is the crucial step and is based on core model results from [13] and on results of K. Prikry and J. Silver from [12]. For some ultrafilter U on κ let $\kappa^\kappa/_U$ denote the ultrapower of κ mod U and for $f \in \kappa^\kappa$ let $[f]_U \in \kappa^\kappa/_U$ denote the equivalence class of f mod U. Let $C_\delta \in \kappa^\kappa$ be the function with constant value δ.

Lemma 4.5: Let U be a ν-indecomposable ultrafilter on κ, ν regular $<\kappa$. Let $\delta < \kappa$ and $cf(\delta) = \nu$. Let $\langle\delta_\eta | \eta < \nu\rangle$ be a strictly increasing seuqence such that $\bigcup_{\eta<\nu} \delta_\eta = \delta$. Then in $\kappa^\kappa/_U$ $\langle[C_{\delta_\eta}] | \eta < \nu\rangle$ converges to $[C_\delta]$.

This result has been proved by J. Silver and can be found in [12] as lemma 6. The next lemma is theorem 5 in [12] and due to K. Prikry.

Lemma 4.6: Let κ be a regular cardinal and U a weakly normal ultrafilter on κ. Let $C \subseteq \kappa^{\kappa}/_{U}$ be closed and unbounded below $[id \upharpoonright \kappa]_U$ Then one of the following conditions holds:

(i) $\{\delta \mid [C_\delta] \in C\} \in U$

(ii) $\{\delta \mid U$ is $cf(\delta)$-decomposable$\} \in U$.

Proof of lemma 4.4. By standard covering lemma arguments β is singular in K and $\beta^+ = (\beta^+)^K$. In [13], it is shown that there is a $\square_\beta$-sequence in K. $\beta^+ = (\beta^+)^K$ implies that the $\square_\beta$-sequence in K is just a $\square_\beta$-sequence, i.e. there exists a sequence $\langle C_\lambda \mid \lambda < \beta^+ \wedge \mathrm{Lim}(\lambda)\rangle$ such that

(i) C_λ is closed in λ

(ii) $C_\lambda \in \lambda \rightarrow cf(\lambda) = \omega$

(iii) $ot(C_\lambda) < \beta$ (ot(X) is the order type of X)

(iv) $\tau \in C_\lambda \rightarrow C_\tau = C_\lambda \cap \tau$.

Now we have to cheque two cases.

Case 1: $\{\lambda \in \beta^+ \mid cf(\lambda) = \omega\} \in U$.

Then by lemma 2.9 (b), U is (ω_1, β^+)-regular and the lemma 4.4 is proved.

Case 2: $\{\lambda \in \beta^+ \mid cf(\lambda) > \omega\} \in U$.

Now assume that the following condition holds.

(i) $\{\delta \mid U$ is $cf(\delta)$-decomposable$\} \in U$.

If for all regular $\nu < \beta$ there is a $\delta < \beta^+$ such that $\nu < cf(\delta)$ and U is $cf(\delta)$-decomposable, then lemma 4.4. is established. So assume (i) and that there is a regular $\nu < \beta$ such that for all $\delta < \beta^+$ such that U is $cf(\delta)$-decomposable $cf(\delta) \leq \nu$. But then $\{\delta \mid cf(\delta) \leq \nu\} \in U$ and by lemma 2.9. (b) U is (ν^+, β^+)-regular and hence δ-decomposable for all $\delta \leq \beta$ such that $\nu^+ \leq cf(\delta)$. But then

$\bigcup\{\nu < \beta \mid \nu$ regular and U ν-decomposable$\} = \beta$ and the lemma is proved.

(Evidently this case cannot occur, assuming (i)).

Now assume that (i) is not true. Then

(ii) $\{\delta \mid Lim(\delta) \wedge U$ is $cf(\delta)$-indecomposable$\} \in U$, since U is weakly normal. Hence U is ν-indecomposable for some regular $\nu < \beta$. The set $X = \{\rho < \beta^+ \mid cf(\rho) = \nu\}$ is stationary in β^+. For $\mu < \beta^+$ define

$X_\mu = \{\rho \in X \mid ot(C_\rho) = \mu\}$.

Now for some $\tau < \beta$ X_τ is stationary in β^+. For each $\lambda < \beta^+$, $Lim(\lambda)$ define $\gamma_\lambda \in C_\lambda$ such that $ot(C_{\gamma_\lambda}) = \nu$ and

$$C'_\lambda = \begin{cases} C_\lambda & \text{for } ot(C_\lambda) \leq \nu \\ C_\lambda \setminus (\gamma_\lambda + 1) & \text{for } ot(C_\lambda) > \nu. \end{cases}$$

Hence $\langle C'_\lambda \mid \lambda < \beta^+ \wedge Lim(\lambda)\rangle$ is a $\square_\beta$-sequence and $X_\tau \cap C'_\lambda = \emptyset$ for all limit cardinals $\lambda < \beta^+$.

Now define $C' = \prod_{\lambda<\beta^+} C'_\lambda/U$ to be the ultraproduct with $C'_\lambda = \lambda$ for nonlimit cardinals $< \beta^+$. Since U is weakly normal, it contains all closed unbounded subsets of β^+ and so only the C'_λ from the modified $\square_\beta$-sequence are important.

$\{\lambda \in \beta^+ \mid cf(\lambda) > \omega\} \in U$ implies, that C' is closed and unbounded below $[id \upharpoonright \beta^+]_U$. Since we assume $\{\delta \mid U$ is $cf(\delta)$-decomposable$\} \notin U$, by lemma 4.6 $D = \{\delta \mid [C_\delta] \in C'\} \in U$. Since X_τ is a stationary set containing only elements of cofinality ν, there is a strictly increasing sequence of type ν of elements of D, $\langle \delta_\eta \mid \eta < \nu\rangle$ such that $\delta = \bigcup_{\eta<\nu} \delta_\eta \in X_\tau$. Now lemma 4.5 implies $\delta \in D \cap X_\tau$. But this shows $\{\lambda < \beta^+ \mid \delta \in C'_\lambda\} \in U$ and contradicts $C'_\lambda \cap X_\tau = \emptyset$. This shows that (ii) is impossible and the proof of lemma 4.4 is complete. The proof of this lemma makes extensive use of ideas of Ketonen from [8].

Let $\mu \leq \kappa$ be cardinals and $LCH_{\mu\kappa}$ be the statement that all limit cardinals $\lambda \leq \kappa$ such that $\mu \leq cf(\lambda) < \lambda$ are strong limit numbers.

Theorem 4.7: Assume $\neg L^{\mu}$. Let U be a uniform ultrafilter on a regular cardinal κ. Assume $LCH_{\omega_2\kappa}$. Then U is δ-decomposable for all $\delta \leq \kappa$.

Theorem 4.7 follows immediately from theorem 4.3 and the following two results.

Lemma 4.8: Assume $\neg L^{\mu}$. Let β be a singular strong limit cardinal and U a uniform ultrafilter on β^+. Then U is (ω,β)-regular.

Lemma 4.9: Let κ,λ be cardinals, $cf(\lambda) < \lambda \leq \kappa$. Assume that U is an ultrafilter on κ such that U is $(\omega, cf(\lambda))$-regular. Let for a sequence $\langle\delta_\nu | \nu < cf(\lambda)\rangle$ of cardinals such that $\bigcup_{\nu<cf(\lambda)} \delta_\nu = \lambda$ U be δ_ν-decomposable for $\nu < cf(\lambda)$. Then U is λ-decomposable.

Proof of Lemma 4.8: Since β is singular, standard covering lemma arguments show $2^\beta = \beta^+$. This implies $\beta^{+\beta^+} = \beta^{+\beta} = \beta^+ \cdot 2^\beta = \beta^+$. Then by theorem 2.8 U cannot have a first function and lemma 4.1 shows that U is (ω,β)-regular.

Proof of lemma 4.9: Let w.l.o.g. $\langle\delta_\nu | \nu < cf(\lambda)\rangle$ be a strictly increasing sequence converging to λ such that for each $\nu < cf(\lambda)$ $\langle p^\nu_\eta | \nu < \delta_\nu\rangle$ is a δ_ν-decomposition for U. Let $\langle U_\nu | \nu < cf(\lambda)\rangle$ be an injective sequence of elements of U such that $\bigcap_{\nu\in S} U_\nu = \emptyset$ for all infinite $S \subset cf(\lambda)$. W.l.o.g. we may assume $\bigcup_{\nu<cf(\lambda)} U_\nu = \kappa$. For $\mu \in \kappa$ let $x_\mu = \{\nu < cf(\lambda) | \mu \in U_\nu\}$. Then $x_\mu \in \mathcal{P}_\omega(cf(\lambda))$ and

$$H = \{y \in \mathcal{P}_\omega(cf(\lambda)) | \exists\mu \in \kappa\ (y = x_\mu)\}$$

has cardinality κ. Let $\langle p_x | x \in H\rangle$ be the partition of κ generated by the mapping $\mu \to x_\mu$ of κ onto H. Now let for each $x \in H$ $\tilde{U}_x$ be the common refinement of the partitions $\langle p^{\nu}_{\eta} | \eta < \delta_{\nu} \rangle$ for $\nu \in X$ and the partition containing the sets p_x and $\kappa \setminus p_x$. Let $\tilde{\tilde{U}}_x = \{y | \emptyset \neq y \subseteq p_x$ and $y \in \tilde{U}_x\}$. Next we show that $U^* = \bigcup_{x \in H} \tilde{\tilde{U}}_x$ is a λ-decomposition for U. Clearly U^* is a partition of κ. Let $Y \subseteq U^*$ be such that $\cup Y \in U$. Then for each $\nu < cf(\lambda)$ $\cup Y \cap U_{\nu} \in U$. Since for each $\mu \in \cup Y \cap U_{\nu}$ $\nu \in x_{\mu}$, μ is an element of a uniquely determined $y \in \tilde{\tilde{U}}_x$ for some x containing ν. Hence $y \subseteq p^{\nu}_{\eta}$ for some $\eta < \delta_{\nu}$ by the definition of U_x, Since $\langle p^{\nu}_{\eta} | \eta < \delta_{\nu} \rangle$ is a δ_{ν}-decomposition, $\{y \in Y | \exists \eta < \delta_{\nu}\ (U_{\nu} \cap y) \subseteq p^{\nu}_{\eta}\}$ has cardinality $\geq \delta_{\nu}$. This shows $\bar{\bar{Y}} \geq \lambda$. Since for $x \in H$ $\bar{\bar{\tilde{\tilde{U}}}}_x \leq \delta_{max(x)} < \lambda$, we obtain $\lambda = \bar{\bar{U}}^* = \bar{\bar{Y}}$ and U^* gives rise to a λ-decomposition for U.

References

[1] J. Baumgartner, Ineffability properties of cardinals II, in Butts and Hintikka (eds.), Logic, Found. of Math. and Comp. Theory, 87-106, D. Reidel (1977)

[2] C.C. Chang and H.J. Keisler, Model Theory, North Holland, Amsterdam (1973)

[3] T. Dodd and R. Jensen, The core model, unpublished

[4] T. Dodd and R. Jensen, The core model, to appear in Ann. Math. Logic

[5] L. Harrington and A. Kechris, Π^1_2 singletons and $0^\#$. Fund. Math. (95), 167-171

[6] A. Kanamori, Weakly normal filters and irregular ultrafilters, Trans. of AMS 220 (1976), 393-399

[7] J. Ketonen, Strong compactness and other cardinal sins, Ann. Math. Logic 5 (1972), 47-76

[8] J. Ketonen, Nonregular ultrafilters and large cardinals, Trans. of AMS 224 (1976), 61-73

[9] K. Kunen and K. Prikry, On descendingly incomplete ultrafilters, Journ. Symb. Logic 36 (1971), 650-693

[10] W. Mitchell, Ramsey cardinals and constructibility, to appear

[11] J. Paris, Patterns of indiscernibles, Bull. London Math. Soc. 6 (1974), 183-188

[12] K. Prikry, On descendingly complete ultrafilters, Cambridge Summer School in Math. Logic, Lecture Notes in Math. Vol. 337, Springer, New York (1973), 459-488

[13] P. Welsh, Combinatorical Principles in the core model, D. Phil. thesis, Oxford (1979).

A LATTICE STRUCTURE ON THE ISOMORPHISM TYPES OF COMPLETE BOOLEAN ALGEBRAS

Sabine Koppelberg[1)]

II. Mathematisches Institut der FU Berlin
Königin-Luise-Str. 24/26
D-1000 Berlin 33

For every complete Boolean algebra (cBA) B, let $\tau(B)$ be the isomorphism type of B. Let $\tau(A) \leq \tau(B)$ if A is, up to isomorphism, a direct factor of B. Let

$$T(B) = \{\tau(B \upharpoonright b) \mid b \in B\}$$

where the relative algebra $B \upharpoonright b = \{x \in B \mid x \leq B\}$ is endowed with the partial order inherited by B and hence a cBA. $(T(B), \leq)$, the "type structure" of B, is a partially ordered set, as proved in [13], 1.31 or [10], 22.6 (see also 1.4), which has a greatest element $\tau(B)$ and a smallest element $\tau(B \upharpoonright 0)$.

Our most general result how the type structure $(T(B), \leq)$ looks like for arbitrary B is Theorem B in section 2: $(T(B), \leq)$ is a distributive lattice; both $(T(B), \leq)$ and its dual lattice $(T(B), \geq)$ are Stone algebras and Heyting algebras satisfying $(a \rightarrow b) \vee (b \rightarrow a) = 1$ for arbitrary a,b. To get Theorem B we prove in Theorem A that $(T(B), \leq)$ is isomorphic to the structure of global sections of a sheaf of linear orders over the Stone space of R(B), where

$$R(B) = \{x \in B \mid f(x) = x \text{ for every automorphism } f \text{ of } B\}$$

---

1) The author gratefully acknowledges partial support by the Forschungsinstitut für Mathematik, ETH Zürich

is the complete subalgebra of invariant elements of B.

Our starting point for having a closer look at type structures was Theorem C in section 3 which is perhaps the most interesting result of this paper. It says that if T(B) is a linear order, then B is a power of a homogeneous cBA (and hence T(B) is well-ordered), thus answering a question of a preliminary version of [3]. Only after finishing a first version of this paper the author became aware of the fact that this theorem is an unpublished result by Solovay and is quoted in section 3.7 of [5]. Conversely Solovay, after having seen the first version of this paper, has obtained considerably stronger results about the structure of T(B). Applying to the situation $R \subseteq B$ (where $R = R(B)$ is the complete subalgebra of B defined above) the technique of two-stage Boolean-valued forcing as developed in [11] and using Theorem C inside of $V^{(R)}$ he shows that T(B) is a complete lattice. The preliminary version of the present paper only established this for Boolean algebras B satisfying the countable chain condition and for algebras $B = C^I$ where C is a rigid cBA. - More precisely, Solovay's method shows that there are families $(R_i)_{i \in I}$ of cBA's and $(\alpha_i)_{i \in I}$ of ordinals such that

$$R \cong \prod_{i \in I} R_i \ , \quad T(B) \cong \prod_{i \in I} (\alpha_i,<)^{(R_i)}$$

where $\mathfrak{A}^{(C)}$ denotes (the two-valued reduct of) the Boolean power of an $\mathcal{L}$-structure $\mathfrak{A}$ w.r.t. a cBA C.

Solovay, in fact, has proved something more: it is well-known that the class $\mathbf{T}$ of all isomorphism types of cBA's, endowed with two operations $\oplus$ and $\bigoplus$, is a cardinal algebra in the sense of [13]. Here, $\tau(A) \oplus \tau(B)$ is defined to be $\tau(A \times B)$; hence the partial order $\leq$ of $\mathbf{T}$ defined above is expressible by

$$s \leq t \quad \text{iff} \quad t = s \oplus r \quad \text{for some} \quad r \in \mathbf{T}.$$

Now, for a fixed cBA B, let $\mathbf{T}_\infty(B)$ be the class of isomorphism types $\tau(E)$, where E is a cBA such that for each $e \in E \setminus \{0\}$, there are $d \in E$, $b \in B$ satisfying $0 < d \leq e$, $E \upharpoonright d \cong B \upharpoonright b$. $\mathbf{T}_\infty(B)$ is then a cardinal subalgebra of $\mathbf{T}$ the structure of which has been completely described by Solovay.

Another consequence of Solovay's method is that $B^\omega \cong B$ for every cBA B without rigid factors. Note that this, together with some elementary remarks in 4.6 and 4.7, gives immediately Theorem E in 4. - The results quoted above will be published by Solovay in a forthcoming paper.

In contrast to that paper, we do not apply the technique of Boolean-valued forcing. To make the paper more readable, we sometimes give proofs of elementary facts to be found in the literature. This applies, for example, to facts connected with the automorphism group of B or with R(B) contained in [7] or [9] and to facts on cardinal algebras contained in [13].

In section 1 we make some simple remarks showing mainly that every product decomposition of R(B) gives rise to a product decomposition of T(B) (this is, of course, the intuitive background for Theorem A). The reader who is mainly interested in sections 3 and 4 may read these immediately after 1.1 to 1.5. Section 2 contains the Theorems A and B about the structure of $(T(B,\leq)$ in general. In section 3 we give the proof of Theorem C mentioned above, and in section 4 two applications of Theorem C: every cBA B can be completely embedded into a homogeneous cBA C such that the Souslin number of C, compared with that of B, is not too large (Theorem D; remember that, by Proposition 2 in [1], $V = L$ implies that there is a cBA B satisfying the countable chain condition such that B cannot be completely embedded into a homogeneous cBA satisfying the countable chain condition). The second application is Theorem E which solves another problem in the first version of [3]:

if B is a cBA such that Aut(B), the automorphism group of B, is infinite, then $|\mathrm{Aut}\ B|^{\omega} = |\mathrm{Aut}\ B|$.

BA's are abbreviated by their underlying sets. The finite operations on a cBA are denoted by $+,\cdot,-,0,1$, the infinite joins and meets by $\sum$ and $\prod$. We write $a \dot{+} b$ instead of $a+b$ if a,b are disjoint and $\dot{\sum}_{i\in I} a_i$ instead of $\sum_{i\in I} a_i$ if the a_i are pairwise disjoint. If $\dot{\sum}_{i\in I} b_i = 1$, we call $(b_i)_{i\in I}$ a partition of B. B is then isomorphic to the product algebra $\prod_{i\in I}(B \upharpoonright b_i)$, and every product decomposition of B arises, up to isomorphism, in this way. Especially, we have that the direct factors of B are, up to isomorphism, exactly the relative algebras $B\upharpoonright b$ of B. A subset D of $B\setminus\{0\}$ is said to be dense in B if, for each $b \in B\setminus\{0\}$, there is some $d \in D$ such that $d \le b$.

We use standard set-theoretical notations; ω is the set of nonnegative integers and $|X|$ is the cardinality of a set X.

The author expresses her gratitude to M.Richter and M.Rubin for helpful comments on the proof of Theorem C - but mostly to R.Solovay for pointing out an error in section 2 and a considerable simplification in Theorem C; moreover for correspondence about his methods improving the results of this paper.

1. Decompositions of R(B) and T(B) into products

For a cBA B, let Aut B be the automorphism group of B. If C is a cBA, B and C are said to be totally different if there are no $b \in B\setminus\{0\}$ and $c \in C\setminus\{0\}$ such that $B\upharpoonright b \cong C\upharpoonright c$. $x,y \in B$ are totally different if $B\upharpoonright x$, $B\upharpoonright y$ are totally different. Let

$$R(B) = \{r \in B \mid f(r) = r \text{ for every } f \in \mathrm{Aut}\ B\},$$

We write R instead of R(B) if B is fixed. Clearly R is a complete subalgebra of B, the algebra of invariant elements of B.

1.1. Let $a \in B$. Then $a \in R$ iff a and $-a$ are totally different: let $a \in R$ and assume that $0 < x \leq a$, $0 < y \leq -a$ and $B \upharpoonright x \cong B \upharpoonright y$. There is an $f \in \operatorname{Aut} B$ interchanging x and y such that $f(z) = z$ for $z \leq -(x+y)$ which shows that $f(a) \neq a$, a contradiction. On the other hand, let a and $-a$ be totally different and $f \in \operatorname{Aut} B$. $f(a) \cdot -a = 0$, since $f(a)$ and $-a$ are totally different. So $f(a) \leq a$ and, by symmetry, $f(a) = a$. -If $(r_i)_{i\in I}$ is a partition of B such that r_i and r_j are totally different for $i \neq j$, then each r_i is totally different from $\sum_{j\neq i} r_j = -r_i$, and hence $\{r_i \mid i \in I\} \subseteq R$.

1.2. For $a \in B$, let

$$\bar{a} = \sum\{f(a) \mid f \in \operatorname{Aut} B\}.$$

It is clear that $a \leq \bar{a}$ and $\bar{a} \in R$. Moreover, if $r \in R$ and $a \leq r$, then $f(a) \leq f(r) = r$ for $f \in \operatorname{Aut} B$ and hence $\bar{a} \leq r$. So $\bar{a}$ is the smallest element of R which is greater or equal to a, i.e.

$$\bar{a} = \prod\{r \in R \mid a \leq r\},$$

and $a \in R$ iff $a = \bar{a}$.

1.3. For $r \in R$ we have $R(B \upharpoonright r) = R \upharpoonright r$: first, suppose $x \in R(B \upharpoonright r)$. So $x \leq r$. Let $f \in \operatorname{Aut} B$; put $g = f \upharpoonright (B \upharpoonright r)$. Since $g \in \operatorname{Aut}(B \upharpoonright r)$, we get $f(x) = g(x) = x$; this proves $x \in R$. To prove the converse, let $x \in R \upharpoonright r \subseteq B \upharpoonright r$. Let $g \in \operatorname{Aut}(B \upharpoonright r)$. Choose $f \in \operatorname{Aut} B$ such that $g \subseteq f$ and $f(y) = y$ for $y \leq -r$ Then $g(x) = f(x) = x$, since $x \in R$. This shows $x \in R(B \upharpoonright r)$.

If $(r_i)_{i\in I}$ is a partition of R, we have $B = \prod_{i\in I} B_i$, where $B_i \cong B \upharpoonright r_i$. Let $R_i = R(B_i)$. So

$$R \cong \prod_{i\in I} (R \upharpoonright r_i) = \prod_{i\in I} R(B \upharpoonright r_i) = \prod_{i\in I} R_i,$$

i.e. each product decomposition $B \cong \prod_{i\in I} B_i$ given by a partition of $R \subseteq B$ gives rise to a product decomposition $R(B) \cong \prod_{i\in I} R(B_i)$.

1.4. For each cBA B, let $\tau(B)$ be the isomorphism type of B. We may assume, by defining $\tau(B)$ to be the set of BA's isomorphic to B which are of minimal set-theoretical rank, that $\tau(B)$ is a set rather than a proper class; hence we may freely speak about sets and even classes of isomorphism types. Let $\mathbf{T}$ be the class of all isomorphism types $\tau(B)$ where B is a cBA. For any family $(t_i)_{i\in I}$ in T, we define $\bigoplus_{i\in I} t_i \in \mathbf{T}$ by

$$\bigoplus_{i\in I} t_i = \tau(\prod_{i\in I} B_i) \text{ where } \tau(B_i) = t_i \text{ for } i \in I.$$

If $I = \{1,\ldots,n\}$, we write $t_1 \oplus \ldots \oplus t_n$ instead of $\bigoplus_{i\in I} t_i$. If α is a cardinal and $t \in \mathbf{T}$, let

$$\alpha \bullet t = \bigoplus_{i\in I} t_i \text{ where } |I| = \alpha \text{ and } t_i = t \text{ for } i \in I.$$

There are some obvious associative, commutative and distributive laws for $\oplus$, $\bigoplus$ and $\bullet$, e.g. $\alpha \bullet (s \oplus t) = (\alpha \bullet s) \oplus (\alpha \bullet t)$, $(\alpha \oplus \beta) \bullet t = (\alpha \bullet t) \oplus (\beta \bullet t)$ etc.

Define, for s and t in $\mathbf{T}$,

$$s \le t \text{ iff } t = s \oplus x \text{ for some } x \in \mathbf{T}.$$

This means that each cBA of type t has a direct factor of type s. For the convenience of the reader, we give the proof that $\le$ is a partial order of $\mathbf{T}$: clearly, $\le$ is reflexive and transitive. Now suppose $s,t \in \mathbf{T}$ and $s \le t$, $t \le s$. Choose $x,y \in \mathbf{T}$ such that $t = x \oplus s$, $s = y \oplus t$. Let B be a cBA of type s. By induction, choose a_n, b_n, c_n, d_n in B such that

$$1 = a_o$$

$$a_n = d_n + b_n$$

$$b_n = c_n + a_{n+1}$$

$$\tau(B \upharpoonright a_n) = s,\ \tau(B \upharpoonright b_n) = t,\ \tau(B \upharpoonright c_n) = x,\ \tau(B \upharpoonright d_n) = y.$$

So $a_o \ge b_o \ge a_1 \ge b_1 \ge \ldots$. Let $e = \prod_{n\in\omega} a_n = \prod_{n\in\omega} b_n$. Now

$$a_o = e \dotplus \sum_{n\in\omega} c_n \dotplus \sum_{n\in\omega} d_n$$

$$b_o = e \dotplus \sum_{n\in\omega} c_n \dotplus \sum_{n\geq 1} d_n$$

$$s = \tau(B \upharpoonright e) \oplus \omega \bullet x \oplus \omega \bullet y$$

$$t = \tau(B \upharpoonright e) + \omega \bullet x \oplus \omega \bullet y$$

and hence $s = t$.

1.5. Let B be a fixed cBA; for $b \in B$ write $\tau(b)$ instead of $\tau(B \upharpoonright b)$. Put

$$T(B) = \{\tau(b) \mid b \in B\}.$$

If B is fixed, abbreviate $T(B)$ by T. Clearly, $T = \{t \in \mathbf{T} \mid t \leq \tau(B)\}$, and T has a greatest element $1_T = \tau(B)$ and a smallest element $0_T = 0_{\mathbf{T}}$.

Assigning $\tau(b) \in T$ to $b \in B$ gives a map $\tau : B \longrightarrow T$ which is clearly onto and order preserving . Now $\tau \upharpoonright R$ is injective: we prove that $a \not\leq b$ in R implies $\tau(a) \not\leq \tau(b)$. Suppose $\tau(a) \leq \tau(b)$. Choose $x \in B$ such that $0 < x \leq a$ but $x \cdot b = 0$. So $x \leq -b$. Since b and $-b$ are totally different, x is totally different from b, hence from a by $\tau(a) \leq \tau(b)$, which contradicts $x \leq a$. - We may therefore identify R wiht a subset of T, which we shall always do in the sequel, and we have the commutative diagram

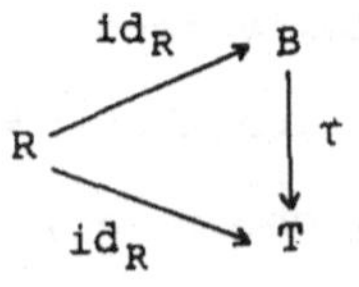

1.6. We may characterize R inside of T in the following way. Although we do not yet know whether $(T,\leq)$ is a lattice, the following are perfectly clear (note that $s,t \in T$ does not imply $s \oplus t \in T$, generally):

(1) for $s,t \in T$, $\inf(s,t) = 0_T$ iff s,t are totally different (i.e., if $a,b \in B$ such that $\tau(a) = s$, $\tau(b) = t$, then a,b are totally different)

(2) if $s,t \in T$ and $\inf(s,t) = 0_T$ then $\sup(s,t)$ exists and equals $s \oplus t$.

Call $t \in T$ complemented if there is $s \in T$ such that $\inf(s,t) = 0_T$ and $\sup(s,t) = s \oplus t = 1_T$. Then

$$R = \{r \in T \mid r \text{ is complemented in } T\},$$

which follows easily from 1.1. It is also not hard to see that every $t \in T$ has a pseudocomplement t^* in T (i.e. a greatest element disjoint from t) which is even in R.

1.7. We proceed to define, for $b \in R$ and $t \in T$, an element $t \wedge b \in T$ (in fact, $t \wedge b$ is the infimum of t and b in $(T,\leq)$, but we will not use this). Namely, let

$$t \wedge b = \tau(x \cdot b) \text{ where } x \in B,\ \tau(x) = t.$$

To show that $t \wedge b$ is well-defined, suppose $y \in B$ such that $\tau(y) = t$. Let $f : B \restriction x \longrightarrow B \restriction y$ be an isomorphism. Now

$$f(x) = f(x \cdot b) \dot{+} f(x \cdot -b) = y \cdot b \dot{+} y \cdot -b,$$

and since $f(x \cdot b)$ is totally different from $y \cdot -b$, $f(x \cdot b) \leq y \cdot b$. The same argument, using f^{-1} instead of f, shows that $f(x \cdot b) = y \cdot b$, hence $\tau(x \cdot b) = \tau(y \cdot b)$.

1.8. For $t \in T$, let

$$T \restriction t = \{s \in T \mid s \leq t\}.$$

So if $b \in B$ and $\tau(b) = t$, $T(B \restriction b) = T \restriction t$. We now prove that each product decomposition of R gives rise to a product decomposition of T (where the product of a family of partially ordered sets is endowed with the coordinate-wise partial order). Let $(b_i)_{i \in I}$ be a partition of R. For $i \in I$, let $B_i = B \restriction b_i$ and $T_i = T(B_i) = T \restriction b_i$. Define

$$\left.\begin{aligned} \varphi : T &\longrightarrow \prod_{i\in I} T_i \\ t &\longmapsto (t \wedge b_i)_{i\in I} \end{aligned}\right\} .$$

Since $t \wedge b_i \leq b_i$ for $i \in I$, $\varphi(t) \in \prod_{i\in I} T_i$ for $t \in T$. φ is one-one: for $t \in T$, let $x \in B$ such that $\tau(x) = t$. By $x = \sum_{i\in I} x\cdot b_i$, we see that $t = \tau(x) = \bigoplus_{i\in I} \tau(x\cdot b_i) = \bigoplus_{i\in I} (t \wedge b_i)$ may be reconstructed from $\varphi(t)$. φ is onto: let $t_i \in T_i$ for $i \in I$, i.e. $t_i \leq b_i$. Choose, for $i \in I$, $x_i \in B$ such that $\tau(x_i) = t_i$ and $x_i \leq b_i \in R \subseteq B$. Put $x = \dot{\sum}_{i\in I} x_i$ and $t = \tau(x)$. For $i \in I$, $t \wedge b_i = \tau(x\cdot b_i) = \tau(x_i) = t_i$; so $\varphi(t) = (t_i)_{i\in I}$. $-\varphi$ is order-preserving: let $s \leq t$. Choose $x, y \in B$ such that $\tau(y) = t$, $x \leq y$ and $\tau(x) = s$. Then, for $i \in I$,

$$s \wedge b_i = \tau(x\cdot b_i) \leq \tau(y\cdot b_i) = t \wedge b_i .$$

Finally, $\varphi(s) \leq \varphi(t)$ (i.e. $s \wedge b_i \leq t \wedge b_i$ for $i \in I$) implies $s \leq t$, since $s = \bigoplus_{i\in I} (s \wedge b_i) \leq \bigoplus_{i\in I} (t \wedge b_i) = t$.

This shows that φ is an isomorphism from T onto $\prod_{i\in I} T(B \upharpoonright b_i)$.

<u>1.9</u>. Define, for our cBA B,

$$a = \sum \{x \in B \mid x \text{ is an atom of } B\},$$

$$h = \sum \{x \in B \mid B \upharpoonright x \text{ is atomless and homogeneous}\}$$

$$r = \sum \{x \in B \mid B \upharpoonright x \text{ is atomless and rigid}\}$$

$$d = -(a+h+r)$$

(a BA C is called rigid if $|\mathrm{Aut}\ C| = 1$ and homogeneous if $C \cong C \upharpoonright c$ for every $c \in C\setminus\{0\}$). a, h, r and hence d are elements of R and pairwise totally different. By 1.8.

$$T = T(B) \cong T(B \upharpoonright a) \times T(B \upharpoonright h) \times T(B \upharpoonright r) \times T(B \upharpoonright d).$$

Now, $B \upharpoonright a$ is isomorphic to the power set algebra $P(\kappa)$ where κ is the cardinality of the set of atoms of B; for $y, z \in P(\kappa)$, $\tau(y) = \tau(z)$ iff $|y| = |z|$. So $T(B \upharpoonright a)$ is isomorphic to the well-ordered set of cardinals $\lambda \leq \kappa$.

Next, (see [7]),

$$B \restriction h \cong \prod_{i \in I} H_i^{\kappa_i}$$

where each H_i is a homogeneous cBA, κ_i is a cardinal and, for $i \neq j$, H_i and H_j are totally different. By 1.8,

$$T(B \restriction h) \cong \prod_{i \in I} T(H_i^{\kappa_i}).$$

If H is a homogeneous atomless cBA and κ a cardinal, $T(H^\kappa)$ is a well-ordered chain with a greatest element (which describes fully the structure of $T(B \restriction h)$): let

$$c(H) = \sup\{|D|^+ | D \subseteq H, \text{ and } d \cdot d' = 0 \text{ for } d, d' \in D, d \neq d'\},$$

the Souslin number of H. It is not hard to prove that, for any cardinals $\alpha, \beta > 0$,

$$H^\alpha \cong H^\beta \text{ iff } \alpha, \beta < c(H).$$

Let $K = \{0,1\} \cup \{\alpha | \alpha \text{ a cardinal, and } c(H) \leq \alpha \leq \kappa\}$. Since, for every $x \in H^\kappa$, $H^\kappa \restriction x \cong H^\alpha$ for exactly one $\alpha \in K$, we have that $(T(H^\kappa), \leq) \cong (K, \leq)$. Again by [7],

$$B \restriction r = \prod_{j \in J} C_j^{\kappa_j}$$

where each C_j is rigid, κ_j is a cardinal and, for $j \neq j'$, C_j and $C_{j'}$ are totally different. It can be proved in an elementary way that, if C is rigid, $T(C^\kappa)$ is a retract of the complete lattice $(K,<)^{(C)}$ and hence itself a complete lattice; here K is the set of cardinals α not greater than κ and $\mathcal{O}^{(C)}$ denotes, for any structure $\mathcal{O}$, the Boolean power of $\mathcal{O}$ by C. If C satisfies the countable chain condition, then $T(C^\kappa) \cong (K, \leq)^{(C)}$. Hence the difficulties in describing T(B) concentrate on $T(B \restriction d)$, where $B \restriction d$ is a cBA without homogeneous or rigid factors - the existence of these algebras was first proved in [1].

1.10. It is easily established that the following conditions on a cBA

B are equivalent:

a) $T = T(B)$ is a Boolean algebra

b) T is a complete Boolean algebra

c) $T = R(B)$

d) $\tau(x) = \tau(\bar{x})$ for $x \in B$

e) each $t \in T$ has a complement in T.

Moreover, for a cBA B satisfying the countable chain condition and without rigid factors, $T(B)$ is always a cBA: it is not hard to see, by the above equivalence, that $T(B)$ is a cBA iff $\omega \circledast t = t$ holds for each $t \in T(B)$. But Solovay's method yields that $\omega \circledast t = t$ for each $t = \tau(A)$ where A is a cBA without rigid factors.

2. The lattice structure of T(B)

We assume the notation of section 1; so $R = R(B)$ is a complete subalgebra of B. Recall that, for $x \in B$, $\bar{x} \in R$ is the smallest element $r \in R$ satisfying $x \leq r$. Let, for the rest of this section,

$$X = \{p \subsetneq R \mid p \text{ is an ultrafilter of } R\},$$

the Stone space of R with the usual topology. We shall define, for $p \in X$, a quotient structure $(T_p, \leq)$ of $T, \leq)$ and a topology on $S = \bigcup_{p \in X} T_p$. We then get a sheaf $\mathcal{S} = (S, \pi, X, \mu)$ of $\mathcal{L}$-structures (as defined in [6]) over X where $\mathcal{L}$ is the language of first-order logic with one binary predicate $<$. We prove:

Theorem A. $(T, <)$ is isomorphic to $(\Gamma(\mathcal{S}), <)$, the $\mathcal{L}$-structure of all global sections of $\mathcal{S}$.

The essential point in this representation of $(T, <)$ is that the stalks $(T_p, <)$ are linear orders and, for $s, t \in T$, $\{p \in X \mid \pi_p(s) < \pi_p(t)\}$ is a clopen subset of X, π_p being the canonical map from T to T_p. This establishes

<u>Theorem B</u>. $(T,<)$ is a distributive lattice with 0 and 1. Both $(T,<)$ and its dual lattice $(T,>)$ are Stone algebras and Heyting algebras. If $\rightarrow$ is the operation of relative pseudo-complementation in $(T,<)$, $(s\rightarrow t)\vee(t\rightarrow s)=1_T$ holds for $s,t\in T$ - i.e. $(T,<)$ is a linear Heyting algebra in the sense of [8].

<u>2.1</u>. In 1.7, we defined $t\wedge b\in T$ for $t\in T$, $b\in R$ by $t\wedge b=\tau(x\cdot b)$ where $\tau(x)=t$. We shall sometimes use the fact that $s\wedge b\leq t\wedge b$ in T and $c\leq b$ in R imply $s\wedge c\leq t\wedge c$: let $x,y\in T$ such that $\tau(x)=s$, $\tau(y)=t$. By $s\wedge b\leq t\wedge b$ there is some $z\in B$ such that $z\leq y\cdot b$ and some isomorphism f from $B\upharpoonright x\cdot b$ onto $B\upharpoonright z$. Now $z=z\cdot c\dotplus z\cdot -c$ and $x\cdot c\leq c$ is totally different from $z\cdot -c\leq -c$. Hence, $f(x\cdot c)\leq z\cdot c\leq y\cdot c$.

<u>2.2</u>. Let $p\in X$ be fixed. For $s,t\in T$ define

$s\leq_p t$ iff there is some $b\in p$ such that $s\wedge b\leq t\wedge b$

$s\sim_p t$ iff $s\leq_p t$ and $t\leq_p s$.

Clearly, $\leq_p$ is a reflexive and transitive relation on T, $\sim_p$ is an equivalence relation and

$s\sim_p t$ iff there is some $b\in p$ such that $s\wedge b=t\wedge b$.

For $s,t\in T$, let

$\pi_p(t)=\{s'\in T\,|\,s'\sim_p t\}$,

$T_p=\{\pi_p(t)\,|\,t\in T\}$

$\pi_p(s)\leq\pi_p(t)$ iff $s\leq_p t$.

So $(T_p,\leq)$ is a partially ordered set, $\pi_p: T\longrightarrow T_p$ is an order-preserving and surjective mapping and T_p has a smallest $\pi_p(O_T)$ and a greatest element $\pi_p(1_T)$.

<u>2.3</u>. Let $s,t\in T$. In the rest of Section 2, we shall frequently refer to the following construction and notations: choose $x,y\in B$ such that

$\tau(x) = s$, $\tau(y) = t$. Let, by Zorn's lemma, $F = \{(x_i, y_i) \mid i \in I\}$ be a family maximal with respect to the properties: $0 < x_i \leq x$, $0 < y_i \leq y$ and $\tau(x_i) = \tau(y_i)$ for $i \in I$; for $i \neq j$, $x_i \cdot x_j = y_i \cdot y_j = 0$. Put

$$x_o = \dot{\sum_{i \in I}} x_i \qquad y_o = \dot{\sum_{i \in I}} y_i$$

$$x_1 = x \cdot -x_o \qquad y_1 = y \cdot -y_o .$$

So

$$\tau(x_o) = \tau(y_o) = d$$

for some $d \in T$ and x_1, y_1 are, by maximality of F, totally different. Hence

$$b = \bar{x}_1, \quad c = \bar{y}_1$$

are disjoint by the definition of $\bar{x}_1, \bar{y}_1$ given in 1.2.

2.4. Let $p \in X$ be fixed. By $b \cdot c = 0$, $b \notin p$ or $c \notin p$. Assume that $b \notin p$. So $f = -b \in p$ and $x_1 \cdot f = 0$. This yields

$$x \cdot f = x_o \cdot f \dot{+} x_1 \cdot f = x_o \cdot f,$$

$$s \wedge f = d \wedge f = \tau(y_o \cdot f) \leq \tau(y \cdot f) = t \wedge f$$

$$s \leq_p t.$$

We have proved that $s \leq_p t$ if $b \notin p$; by symmetry, $t \leq_p s$ follows from $c \notin p$. So for any $p \in X$, $(T_p, \leq)$ is a linear order.

Note that, although $x, y, F, x_o, x_1, y_o, y_1$ are not uniquely determined by s, t, the fact that $s \leq_p t$ or $t \leq_p s$ for every $p \in X$ does, of course, not depend on the choice of $x, y, F, \ldots$. Moreover for each $p \in X$, the type $d \in T$ defined in 2.3 satisfies

$$\pi_p(d) = \min(\pi_p(s), \pi_p(t));$$

this is proved for the case $b \notin p$ by the above lines, and it follows by symmetry that $c \notin p$ implies $d \wedge -c = t \wedge -c$.

2.5. Let $s,t \in T$ be fixed and $b,c \in R$ as in 2.3. Put

$$u = \{p \in X \mid s \leq_p t\}.$$

By the definition of $\leq_p$, it is clear that u is an open subset of X. We want to show that u is even clopen. First, abbreviating the clopen subset $\{p \in X \mid f \in p\}$ of X for $f \in R$ by cl(f), we see that $s \wedge f \leq t \wedge f$ implies $cl(f) \subseteq u$. Let

$$M = \{a \in R \mid s \wedge a \leq t \wedge a\}, \quad \alpha = \textstyle\sum^R M.$$

By 2.1, there is a subset $\{a_i \mid i \in I\}$ of M such that $\alpha = \sum_{i \in I} a_i$. It follows as in 2.1 that $s \wedge \alpha \leq t \wedge \alpha$; i.e. that α is the largest element of M. We claim that $u = cl(\alpha)$. Clearly, by $\alpha \in M$, $cl(\alpha) \subseteq u$. Now suppose $p \in u$. By $p \in u$, choose $f \in p$ satisfying $s \wedge f \leq t \wedge f$. So $f \in M$, $f \leq \alpha$ and $p \in cl(f) \subseteq cl(\alpha)$, which proves $u \subseteq cl(\alpha)$. In an analoguous way, we see that $\{p \in X \mid \pi_p(t) \leq \pi_p(s)\}$ and $\{p \in X \mid \pi_p(s) = \pi_p(t)\}$ are clopen subsets of X.

2.6. W.l.o.g., suppose that for $p \neq q$ in X, $T_p \cap T_q = \emptyset$. Put

$$S = \dot{\bigcup_{p \in X}} T_p.$$

Let $\pi : S \longrightarrow X$ be defined by

$$\pi(x) = p \quad \text{iff} \quad x \in T_p.$$

For $t \in T$, $g \in R$ let

$$b_{t,g} = \{\pi_p(t) \mid p \in cl(g)\}.$$

Then $\{b_{t,g} \mid t \in T, g \in R\}$ is a basis of a topology of S. It is easily checked that π is, w.r.t. this topology, a continuous open map from S onto X. For $x \in S$ there is a neighbourhood n of x in S which is mapped by π homeomorphically onto an open subset of X: let, if $x = \pi_p(t)$, $n = b_{t,1_R}$; $\pi \upharpoonright n$ is an injective continuous open map from n onto X.

2.7. Let μ be the map which assigns to every $p \in X$ the $\mathcal{L}$-structure

$(T_p,<)$. The underlying set of $\mu(p)$ is $T_p = \pi^{-1}(p)$. Suppose $x,y \in T_p$ and $x < y$ in $\mu(p) = (T_p,<)$; e.g. $x = \pi_p(s)$ and $y = \pi_p(t)$ where $s,t \in t$. Then there are neighbourhoods $b_{s,g}$ of x and $b_{t,g}$ of y such that for every $q \in cl(g)$, $\pi_q(s) < \pi_q(t)$ as follows from 2.5.

2.8. The details proved in 2.6 and 2.7 show that $\mathcal{S} = (S,\pi, X,\mu)$ is a sheaf of $\mathcal{L}$-structures over X in the sense of [6]. Recall that a global section of $\mathcal{S}$ is a continuous map $\sigma : X \longrightarrow S$ such that $\pi \circ \sigma = id_X$, i.e. $\sigma(p) \in T_p$ for $p \in X$. $\Gamma(\mathcal{S})$ denotes the set of all global sections of $\mathcal{S}$. Since $\Gamma(\mathcal{S}) \subseteq \prod_{p\in X} T_p$, we may consider $\Gamma(\mathcal{S})$ as a substructure of the product structure $\prod_{p\in X}(T_p,<)$; so $\Gamma(\mathcal{S})$ is (if non-empty) an $\mathcal{L}$-structure. In 2.9 we give examples of global sections of $\mathcal{S}$.

2.9. For $t \in T$, let $\sigma_t : X \longrightarrow S$ be defined by

$$\sigma_t(p) = \pi_p(t).$$

We show that $\sigma_t \in \Gamma(\mathcal{S})$, i.e. that σ_t is a continuous map: let $\sigma_t(p) = x = \pi_p(t)$; let $b_{s,g}$ be a neighbourhood of x. So $g \in p$ and $x = \pi_q(s)$ for some $q \in X$ satisfying $g \in q$. By $\pi_p(t) = \pi_q(s)$ and $T_p \cap T_q = \emptyset$ for $p \neq q$ we get $p = q$ and $\pi_p(t) = \pi_p(s)$. Let, by 2.5,

$$v = \{p' \in X \mid \pi_{p'}(s) = \pi_{p'}(t)\} = cl(\beta).$$

So $p \in v$ and $g \cdot \beta \in p$. We prove that σ_t maps the neighbourhood $cl(g\cdot\beta)$ of p in X into the neighbourhood $b_{s,g}$ of x in S: if $g \cdot \beta \in p'$, $p' \in v$ and

$$\sigma_t(p') = \pi_{p'}(t) = \pi_{p'}(s) \in b_{s,g}$$

since $g \in p'$.

2.10. Proof of Theorem A. We prove that

$$\left.\begin{array}{l} \varphi : T \longrightarrow \Gamma(\mathcal{S}) \\ t \longmapsto \sigma_t \end{array}\right\}$$

is an isomorphism of $\mathcal{L}$-structures. Since $\varphi(t) \in \Gamma(\mathcal{S})$ for $t \in T$, $\Gamma(\mathcal{S}) \neq \emptyset$. Now

$$s \le t \ \text{ in } T \ \text{ iff } \ \sigma_s \le \sigma_t \ \text{ in } \ \Gamma(\mathcal{S}):$$

if $s \le t$, $\sigma_s(p) = \pi_p(s) \le \pi_p(t) = \sigma_t(p)$ holds for every $p \in X$; hence $\sigma_s \le \sigma_t$. Conversely, assume that $\sigma_s \le \sigma_t$. This means, in the terminology of 2.5, that $u = X$, $\alpha = 1$ and, by $\alpha \in M$, $s \le t$. - So φ is one-one and both φ and φ^{-1} are order-preserving; it remains to prove that φ is onto. Let $\sigma \in \Gamma(X,S)$ be given. For $p \in X$ choose $t_p \in T$ such that $\sigma(p) = \pi_p(t_p)$. Since also $\sigma_{t_p}(p) = \pi_p(t_p) = \sigma(p)$, there is an open neighbourhood u_p of p such that $\sigma \restriction u_p = \sigma_{t_p} \restriction u_p$. By compactness of X, choose $b_1,\dots,b_n \in R$ such that $1_R = b_1 \dotplus \dots \dotplus b_n$ and for $1 \le i \le n$ there exists $p_i \in X$ satisfying $cl(b_i) \subseteq u_{p_i}$. Since $b_1,\dots,b_n \in R$ are pairwise disjoint, the isomorphism type

$$t = (t_{p_1} \wedge b_1) \oplus \dots \oplus (t_{p_n} \wedge b_n)$$

is in T. We claim that $\sigma = \sigma_t$. In fact, for $p \in X$, e.g. $b_i \in p$, we have that $p \in u_{p_i}$ and $\sigma(p) = \sigma_{t_{p_i}}(p)$. On the other hand, $t \wedge b_i = t_{p_i} \wedge b_i$ and $\sigma_t(p) = \pi_p(t) = \pi_p(t_{p_i}) = \sigma_{t_{p_i}}(p)$.

2.11. Proof of Theorem B. We prove that $(\Gamma(\mathcal{S}),<)$ is a lattice with the properties claimed in Theorem B. $\varphi(0_T)$ and $\varphi(1_T)$ are, of course, the smallest resp. the largest elements of $\Gamma(\mathcal{S})$. Let σ_s and σ_t be elements of $\Gamma(\mathcal{S})$. Define

$$u = \{p \in X \mid \pi_p(s) \le \pi_p(t)\}, \ v = \{p \in X \mid \pi_p(t) \le \pi_p(s)\}.$$

So u and v are clopen subsets of X. Clearly, the function $\sigma_s \wedge \sigma_t \in \Gamma(\mathcal{S})$ which coincides with σ_s on u and with σ_t on v is the infimum of σ_s and σ_t; likewise, the function $\sigma_s \vee \sigma_t \in \Gamma(\mathcal{S})$ which coincides with σ_t on u and with σ_s on v is the supremum of σ_s and σ_t; the function $\sigma_s \to \sigma_t \in \Gamma(\mathcal{S})$ which coincides with σ_{1_T} on u and with σ_t on $v \setminus u$ is the relative pseudocomplement of σ_s w.r.t. σ_t. $\Gamma(\mathcal{S})$ is a distributive lattice since every $(T_p,<)$, being a linear order, is a distributive lattice and $\sigma_s \wedge \sigma_t$,

$\sigma_s \vee \sigma_t$ are computed coordinate-wise. The equation $(a \to b) \vee (b \to a) = 1$ holds in every $(T_p,<)$ and hence in $\Gamma(\mathcal{S})$. Moreover, $\Gamma(\mathcal{S})$ is a Stone algebra: for $\sigma_t \in \Gamma(\mathcal{S})$, $w = \{p \in X \mid \sigma_t(p) = 0_{T_p}\}$ is a clopen subset of X. Let $(\sigma_t)^* \in \Gamma(\mathcal{S})$ such that $(\sigma_t)^*(p) = 1$ if $p \in w$ and $(\sigma_t)^*(p) = 0$ if $p \notin w$. $(\sigma_t)^*$ is the pseudo-complement of σ_t in $\Gamma(\mathcal{S})$ and $(\sigma_t)^* \vee (\sigma_t)^{**} = \sigma_{1_T}$. The dual lattice of $(\Gamma(\mathcal{S}),<)$ is $(\Gamma(\mathcal{S}),>)$, which is, up to isomorphism, the structure of global sections of the sheaf (X,π,S,μ') where $\mu'(p) = (T_p,>)$ if $\mu(p) = (T_p,<)$. So the same arguments show that $(\Gamma(\mathcal{S}),>)$ is a Stone algebra and a Heyting algebra.

__2.12.__ Let $(s_i)_{i\in I}$ be a family in T such that $\bigvee_{i\in I} s_i$ (the supremum of the s_i in T) exists. For $t \in T$, $\bigvee_{i\in I}(t \wedge s_i)$ also exists and

$$t \wedge \bigvee_{i\in I} s_i = \bigvee_{i\in I}(t \wedge s_i),$$

since $(\Gamma(\mathcal{S}),\vee,\wedge,\to)$ is a Heyting algebra. By duality,

$$t \vee \bigwedge_{i\in I} s_i = \bigwedge_{i\in I}(t \vee s_i)$$

if $\bigwedge_{i\in I} s_i$ exists. Of course, no distributive laws of the form $\bigwedge_{i\in I}\bigvee_{j\in J} t_{ij} = \bigvee_{f\in J^I}\bigwedge_{i\in I} t_{if(i)}$ for $|I| \geq \omega$, $|J| \geq 2$ can be proved for T, since R is a complete sublattice of T (R is the set of complemented elements of T, as was already noticed in 1.6) and these laws do not hold in arbitrary cBA's.

__2.13.__ Let $s,t \in T$ and choose $x,y,\dots$ as in 2.3. Then $d = \tau(x_o) = \tau(y_o)$ is the infimum of s and t in T which follows by 2.4 and the description of $\sigma_s \wedge \sigma_t$ given in 2.11. Thus, if $\tau(x) = s$, $\tau(y) = t$, $x_o \leq x$ and $y_o \leq y$ are chosen such that $\tau(x_o) = \tau(y_o)$ but $x \cdot - x_o$, $y \cdot - y_o$ are totally different, then $\tau(x_o) = s \wedge t$. This yields e.g. the following fact:
let $A \cong B \times C \cong D \times E$, $A' \cong B' \times C' \cong D' \times E'$ where A,A' (and hence B,B'...) are cBA's and suppose $B \cong B'$, $D \cong D'$, C and C' resp. E and E' are totally different. Then $B \cong D$, since $\tau(B) = \tau(A) \wedge \tau(A') = \tau(D)$.

2.14. Let $(V,\wedge,\vee,^*)$ be a Stone algebra and $Sk = \{x^* | x \in V\}$. Then Sk is a subalgebra of V and a Boolean algebra which is called the skeleton of V in [4]. Let X be the Stone space of Sk. For $p \in X$ and $x,y \in V$, put $x \leq_p y$ iff $x \wedge b \leq y \wedge b$ for some $b \in p$. We may repeat the whole construction described in 2.2, 2.6 to 2.10 to see that $\pi_p : V \to V/{\sim_p} = V_p$ is an epimorphism of Stone algebras and $V \cong \Gamma(\mathfrak{F}_c)$ where $\mathfrak{F}_c$ is the "canonical" sheaf over X with stalks V_p. It is not hard to prove that the following conditions are equivalent.

(a) both V and its dual lattice are Stone algebras and V is a Heyting algebra satisfying $(a\to b) \vee (b\to a) = 1$ for $a,b \in V$;

(b) every stalk V_p of $\mathfrak{F}_c$ is a linear order and, for $\sigma,\sigma' \in \Gamma(\mathfrak{F}_c)$, $\{p \in X | \sigma(p) = \sigma'(p)\}$ is a clopen subset of X (i.e. $S_c = \dot{\bigcup}_{p\in X} V_p$ is a Hausdorff space).

Of course, not every Stone algebra satisfying (a) can have, up to isomorphism, the form T(B) for some cBA B: $Sk(T(B)) = R(B)$ is a complete BA and, if R(B) is finite, every stalk T_p of $\mathfrak{F}$ is well-ordered by Theorem C and 1.9.

3. Weakly homogeneous cBA's

In this section we define weak homogeneity, a property of certain cBA's. Our main goal is

Theorem C (Solovay). Every weakly homogeneous cBA is, up to isomorphism, a power of a homogeneous cBA.

We first give in 3.1 several equivalent formulations of weak homogeneity and in 3.2 equivalences to being a power of a homogeneous cBA, although some of these equivalences may be known to the reader.

3.1. For a cBA, let $R = R(B)$ and $T = T(B)$ be defined as in Section 1. Then the following conditions are equivalent and a cBA satisfying any of them will be called weakly homogeneous:

a) $R = \{0,1\}$

b) if $a,b \in B\setminus\{0\}$, there are $x,y \in B$ such that $0 < x \leq a$, $0 < y \leq b$, and $B \upharpoonright x \cong B \upharpoonright y$ (i.e. a,b are not totally different)

c) $(T,\leq)$ is a linear order

d) if $s,t \in T\setminus\{0_T\}$, there exists $r \in T$ such that $0 < r \leq s,t$.

Proof. a→b: if $0 < a,b \in B$ and a,b are totally different, then $0 < a \leq \bar{a} \in R$, $0 < b \leq \bar{b} \in R$ and $\bar{a}\,\bar{b} = 0$, hence R has at least four elements.

b) → c): let $a,b \in B$. By Zorn's lemma, let $F = \{(x_i,y_i) \mid i \in I\}$ be a maximal family w.r.t. the conditions: $0 < x_i \leq a$, $0 < y_i \leq b$, $\tau(x_i) = \tau(y_i)$, and $i \neq j$ implies $x_i \cdot x_j = y_i \cdot y_j = 0$. By b) and the maximality of F, $\dot{\sum\limits_{i\in I}} x_i = a$ or $\dot{\sum\limits_{i\in I}} y_i = b$. Assume $\dot{\sum\limits_{i\in I}} x_i = a$. Then

$$\tau(a) = \bigoplus_{i\in I}\tau(x_i) = \bigoplus_{i\in I}\tau(y_i) \leq \tau(b).$$

c) → d) is trivial.

d) → a): suppose $r \in R$ and $0 < r < 1$. Then r and -r are totally different and, being elements of T, they satisfy $\inf_T(r,-r) = 0_T$.

3.2. On the other hand, the following conditions are equivalent:

e) $B \cong H^\kappa$ where H is a homogeneous cBA and κ a cardinal

f) $(T,\leq)$ is well ordered

g) If $|T| \geq 2$, then T has a least positive element t_o.

Proof. e) → f) was shown in 1.9. f) → g) is trivial.

g) → e): let $t_o \in T$ as in g). Let H be a cBA such that $\tau(H) = t_o$, so H is homogeneous. Now, by g),

$$D = \{x \in B\setminus\{0\} \mid \tau(x) = t_o\}$$

is a dense subset of B. Hence there is a partition $(x_\alpha)_{\alpha<\kappa}$ of B for some cardinal κ such that $x_\alpha \in D$ for $\alpha < \kappa$, and $B \cong \prod\limits_{\alpha<\kappa} B\upharpoonright x_\alpha \cong H^\kappa$.

3.3. Trivially, each power of a homogeneous cBA is weakly homogeneous. We begin with the proof of the converse, i.e. Theorem C. So we assume in 3.3 to 3.5 that B is weakly homogeneous but not a power of a homogeneous BA, i.e. that

(*) $(T,\leq)$ is linearly ordered, but $T^+ = T\setminus\{0_T\}$ has no smallest element,

and we want to get a contradiction. Since (*) holds for every relative algebra $B\restriction b$, where $b>0$, as well, we shall frequently pass to a suitable $B\restriction b$ or assume w.l.o.g. that $b=1$. Our proof will split up in two cases. In case A (3.4), a certain cardinal invariant δ of B will be uncountable and the proof will be easy. In case B (3.5), where $\delta=\omega$, it turns out that T is isomorphic to the closed unit interval of real numbers and $\tau : B \longrightarrow T$ gives rise to a strictly positive σ-additive measure; hence B is a measure algebra and homogeneous.

We shall use in 3.5 the fact that $s \oplus t = t$ for $s,t \in \mathbf{T}$ implies $\omega \cdot s \leq t$: choose a cBA C of type t. By $s \oplus t = t$, we may choose a sequence $(c_n)_{n\in\omega}$ in C satisfying

$$1 = c_o \geq c_1 \geq c_2 \geq \ldots$$

$$\tau(c_n) = t \qquad \tau(c_n - c_{n+1}) = s$$

This shows that

$$\omega \odot s = \tau\Big(\sum_{n\in\omega} c_n - c_{n+1}\Big) \leq \tau(C) = t.$$

Call $t \in \mathbf{T}$ α-divisible (where α is a cardinal) if there is some $s \in \mathbf{T}$ such that $\alpha \odot s = t$ and almost α-divisible if there is some $s \in \mathbf{T}$, $s \neq 0_{\mathbf{T}}$, such that $\alpha \odot s \leq t$. If $\alpha \geq \omega$ and t is α-divisible, then $\alpha \odot t = t$, for $t = \alpha \odot s$ implies $\alpha \odot t = \alpha \odot (\alpha \odot s) = \alpha \odot s = t$. If every $t' \leq t$ is almost α-divisible, then t is α-divisible: let C be a cBA of type t. By assumption,

$$D = \{x \in C\setminus\{0\} \mid \tau(x) = \alpha \odot s \text{ for some } s \in \mathbf{T}\}$$

is a dense subset of C. Let $(c_i)_{i\in I}$ be a partition of C such that $\{c_i \mid i \in I\} \subseteq D$; suppose $\tau(c_i) = \alpha \odot s_i$. Then

$$t = \tau(C) = \bigoplus_{i \in I} \tau(c_i) = \bigoplus_{i \in I} \alpha \circledast s_i = \alpha \circledast \bigoplus_{i \in I} s_i .$$

Finally, if C is a cBA without rigid factors, $n \in \omega \setminus \{0\}$ and $c \in C$, then $\tau(c)$ is almost n-divisible and hence, by the foregoing argument, n-divisible: let $c \in C \setminus \{0\}$. Since $C \upharpoonright c$ is not rigid, there are disjoint $d_1, d_2 \leq c$ such that $\tau(d_1) = \tau(d_2)$ and $d_1, d_2 \neq 0$. Let $k \in \omega$ such that $n \leq 2^k$. Applying the above splitting device k times, we get 2^k pairwise disjoint elements $e_1, \ldots, e_{2^k} \leq c$ which are non-zero and have the same type, say s. Hence $n \circledast s \leq 2^k \circledast s \leq \tau(c)$.

Now, for the proof of Theorem C, let δ be the smallest cardinal such that some $t_o \in T$ is not almost δ-divisible. Surely $\delta \geq \omega$, since B has no rigid factor. We may assume $\tau(B) = t_o$; otherwise, pass to $B \upharpoonright b$ where $\tau(b) = t_o$. So, for each $b \in B \setminus \{0\}$ $\tau(b)$ is α-divisible if $\alpha < \delta$ but no $t \in T$ is δ-divisible. Moreover, if $0 < \alpha < \delta$ and $\omega_1 \leq \delta$, then $\alpha \circledast \tau(b) = \tau(b)$: This wa proved above for $\alpha \geq \omega$, and for $\alpha < \omega$ we have $\omega < \omega_1 \leq \delta$ and

$$\tau(b) \leq \alpha \circledast \tau(b) \leq \omega \circledast \tau(b) = \tau(b).$$

3.4. Case A: $\omega_1 \leq \delta$. Let $t \in T^+$; we show that $t = 1_T$, contradicting (*) in 3.3. Choose, by Zorn's lemma, a maximal family $(b_i)_{i \in I}$ of pairwise disjoint elements of B satisfying $\tau(b_i) = t$. Put $b = -\sum_{i \in I} b_i$. By maximality, $\tau(b) < t$. By definition of δ, $\alpha = |I|$ is smaller than δ. This yields

$$1_T = \alpha \circledast t \oplus \tau(b) \leq (\alpha+1) \circledast t = t,$$

since $\alpha + 1 \leq \max(\alpha, \omega) < \delta$.

3.5. Case B: $\delta = \omega$. We start out with several simple remarks: if $s, t \in T$ and $s \oplus t = t$, then $s = 0_T$ (for $s \neq 0_T$ would imply $\omega \circledast s \leq t$, contradicting $\delta = \omega$). Hence, for $a < b$ in B, $\tau(a) < \tau(b)$. Furthermore, it is easily checked that

$t \oplus s = t \oplus s'$ implies $s = s'$

if $r < s < t$ in T; $a, c \in B$ and $a < c$, $\tau(a) = r$, $\tau(c) = t$, then there exists $b \in B$ such that $a < b < c$ and $\tau(b) = s$

if $t = n \circledast s = n \circledast s'$ and $n \in \omega \setminus \{0\}$, then $s = s'$.

Note that the last assertion is proved in [13] as 2.34 under much more general conditions but with a much longer proof. The unique $s \in T$

satisfying $n \circledast s = t$ will be denoted by $\frac{1}{n} \circledast t$. Denote by Q the set of rational numbers q satisfying $0 \le q \le 1$. For $q = \frac{k}{n} \in Q$ and $t \in T$, let

$$q \circledast t = k \circledast (\tfrac{1}{n} \circledast t).$$

Then

$$(q+q') \circledast t = (q \circledast t) \oplus (q' \circledast t) \text{ if } q+q' \in Q$$

$$q \circledast t \le q' \circledast t \text{ in } T \text{ iff } q \le q' \text{ in } Q$$

etc.

(1) For $0_T < t \in T$, there is some $n \in \omega$ such that $0_T < \frac{1}{n+1} \circledast 1_T < t$: let $X = \{b_1, \ldots, b_n\}$ be a maximal disjoint subset of B satisfying $\tau(b_i) = t$ for $1 \le i \le n$ (X has to be finite by $\delta = \omega$). So $(n \circledast t) \oplus d = 1_T$ where $d = \tau(-(b_1 \dot{+} \ldots \dot{+} b_n))$ and, by maximality, $t \not\le d$. If $t \le \frac{1}{n+1} \circledast 1_T$, we would get $(n+1) \circledast t \le 1_T = (n \circledast t) \oplus d$ and we could conclude $t \le d$. So $\frac{1}{n+1} \circledast 1_T < t$.

(2) For $s < t$ in T, there is $q \in Q$ such that $s < q \circledast 1_T < t$: let $t = s \oplus d$ for some $d \in T$. By (1), let $\frac{1}{n} \circledast 1_T < d$ where $n \in \omega \setminus \{0\}$. Choose $k \in \{0,1,\ldots,n-1\}$ maximal w.r.t. $\frac{k}{n} \circledast 1_T \le s$. So $s < \frac{k+1}{n} \circledast 1_T$; we prove that $\frac{k+1}{n} \circledast 1_T < t$: otherwise, let $\frac{k}{n} \circledast 1_T \oplus d' = s$, $t \oplus d'' = \frac{k+1}{n} \circledast 1_T$. It follows that $d' \oplus d \oplus d'' = \frac{1}{n} \circledast 1_T$ and $d \le \frac{1}{n} \circledast 1_T$, a contradiction.

Define $\varphi : T \to [0,1] = \{x \in \mathbb{R} \mid 0 \le x \le 1\}$ by

$$\varphi(t) = \sup\{q \in Q \mid q \circledast 1_T \le t\}.$$

Clearly φ is order-preserving and $\varphi(q \circledast 1_T) = q$ for $q \in Q$. By (2), φ is injective. Moreover, $\varphi(s \oplus t) = \varphi(s) + \varphi(t)$ if $s \oplus t \in T$ - this holds for $s,t \in \{q \circledast 1_T \mid q \in Q\}$ and follows by monotonicity and the fact that $\{q \circledast 1_T \mid q \in Q\}$ and Q are dense in T resp. $[0,1]$ for arbitrary $s,t \in T$. Define $\mu : B \to [0,1]$ by $\mu = \varphi \circ \tau$:

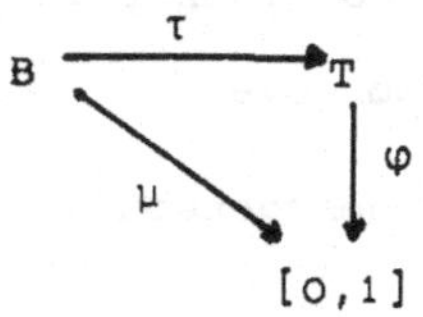

For $0 < b$ in B, $0 < \mu(b)$ follows by (1). So μ is a positive finitely additive measure on B, and

(3) $\tau(b) = \tau(c)$ iff $\mu(b) = \mu(c)$

since φ was one-one.

We show that μ is σ-additive: let $(a_n)_{n\in\omega}$ be a sequence of pairwise disjoint elements of B and $a = \dot{\sum} a_n$. Let $\alpha_n = \mu(a_n)$; we have to prove that $\mu(a) = \sup_{n\in\omega}(\alpha_o + \ldots + \alpha_n)$. For every n,

$$\alpha_o + \ldots + \alpha_n = \mu(a_o + \ldots + a_n) \leq \mu(a).$$

Let $\sigma \in [0,1]$ such that $\alpha_o + \ldots + \alpha_n \leq \sigma$ for every $n \in \omega$; we show that $\mu(a) \leq \sigma$. If not, we may assume $\sigma \in Q$. Choose $c \in B$ such that $\mu(c) = \sigma$. By induction, choose pairwise disjoint $b_n \leq c$ satisfying $\mu(b_n) = \alpha_n$: first, by $\mu(a_o) = \alpha_o \leq \sigma = \mu(c)$, we get $\tau(a_o) \leq \tau(c)$. So there is some $b_o \leq c$ such that $\tau(b_o) = \tau(a_o)$ and $\mu(b_o) = \alpha_o$. Using $c \cdot\!- b_o$, $\sigma - \alpha_o$, α_1, a_1 instead of c, σ, α_o, a_o and $\alpha_o + \alpha_1 \leq \sigma$ we get $b_1 \leq c \cdot\!- b_o$ such that $\mu(b_1) = \alpha_1$ etc. Now let $b = \dot{\sum}_{n\in\omega} b_n$. Then

$$\tau(a) = \bigoplus_{n\in\omega} \tau(a_n) = \bigoplus_{n\in\omega} \tau(b_n) = \tau(b) \leq \tau(c)$$

and $\mu(a) \leq \mu(c) = \sigma$.

We have thus proved that B is a cBA carrying a strictly positive σ-additive measure. The structure theorem in [14] for these algebras says that B is, up to isomorphism, a product of a finite or countable family of pairwise totally different homogeneous measure algebras. Since B is assumed in Theorem C to be weakly homogeneous, B is even homogeneous, hence $T^+ = \{1_T\}$ which again contradicts (*) in 3.3. This concludes Case B and proves Theorem C.

<u>3.6. Corollary.</u> Let $r \in R = R(B)$, $r > 0$. Then r is an atom of R iff $B \upharpoonright r$ is a power of a homogeneous cBA. Moreover

$$a = \sum\{x \in B \mid B \upharpoonright x \text{ is homogeneous}\}$$

is the supremum of all atoms of R.

Proof. By Theorem C, we know that conditions a) through d) and e) through g) of 3.1 and 3.2 are equivalent. Now,

r is an atom of R iff $|R(B \upharpoonright r)| = 2$ (by 1.3)

iff $B \upharpoonright r$ is weakly homogeneous

iff $B \upharpoonright r$ is a power of a homogeneous cBA

by Theorem C. The second assertion follows immediately.

4. Some applications of Theorem C

4.1 Theorem. Let X be an arbitrary topological space and I an infinite set. Then the algebra $C = RO(X^I)$ of regular open subsets of the product space X^I is homogeneous. If α is a cardinal such that $|U| \leq \alpha$ for every set U of pairwise disjoint open subsets of X, then $|V| \leq 2^\alpha$ for every set V of pairwise disjoint open subsets of X^I (so, if $c(RO(X)) \leq \alpha^+$, then $c(RO(X^I)) \leq (2^\alpha)^+$).

Proof. We first prove that C is weakly homogeneous. Let u,v be non-void regular open subsets of X^I. Choose non-void basic open subsets $b \subsetneq u$, $c \subsetneq v$. E.g.

$$b = \prod_{i \notin I_o} X_i \times \prod_{i \in I_o} u_i$$

$$c = \prod_{j \notin J_o} X_j \times \prod_{j \in J_o} v_j$$

where $X_i = X$ for $i \in I$, I_o and J_o are finite subsets of I and u_i, v_j are non-void open subsets of X for $i \in I_o$, $j \in J_o$. We may assume that u_i, v_j are regular open. Let $I_o = \{i_1, \dots, i_n\}$, $J_o = \{j_1, \dots, j_m\}$. Choose $I_1, J_1 \subsetneq I$ such that $I_o \cap I_1 = \emptyset = J_o \cap J_1$, $I_1 = \{i'_1, \dots, i'_m\}$, $J_1 = \{j'_1, \dots, j'_n\}$. Let $u_{i'_k} = v_{j_k}$ for $i \leq k \leq m$, $v_{j'_k} = u_{i_k}$ for $1 \leq k \leq n$. Now,

$$d = \prod_{i \notin (I_o \cup I_1)} X_i \times \prod_{i \in I_o} u_i \times \prod_{i \in I_1} u_i,$$

and

$$e = \prod_{j \notin (J_o \cup J_1)} X_j \times \prod_{j \in J_o} v_j \times \prod_{j \in J_1} v_j,$$

are regular open homeomorphic subsets of X^I such that $d \subsetneq b \subsetneq u$, $e \subsetneq c \subsetneq v$. Hence,

$$C \upharpoonright d \cong RO(d) \cong RO(e) \cong C \upharpoonright e.$$

By Theorem C, C is a power of a homogeneous cBA. To show that C is even homogeneous it is sufficient to show that $c(C \upharpoonright u) = c(C)$ for each $u \in C \setminus \{O\}$ (cf. 1.9). But u contains a non-void basic open subset b and b is homeomorphic to $v \times X^I$ for some open $v \subsetneq X^n$ and some $n \in \omega$; so $c(RO(X^I)) \le c(C \upharpoonright b) \le c(C \upharpoonright u) \le c(RO(X^I))$. The second assertion of the theorem follows by the Erdös-Rado-theorem; see 3.13 in [2].

<u>4.2. Theorem D</u>. Let B be a cBA and let C be the completion of a free product of λ copies of B, where λ is an infinite cardinal. Then C is homogeneous and B is completely embeddable into C. If α is a cardinal such that $c(B) \le \alpha^+$, we have $c(C) \le (2^\alpha)^+$ and $|C| \le (\lambda \cdot |B|)^{2^\alpha}$.

<u>Proof</u>. Let X be the Stone space of B. Then C is isomorphic to $RO(X^\lambda)$ and the theorem follows by 4.1 and the fact that C is completely generated by the λ copies of B.

<u>4.3</u>. Compare this with the example in [1] showing that, if $V = L$ is assumed, there is a cBA B satisfying ccc (the countable chain condition such that, if B is completely embedded into a homogeneous cBA C, $c(C) \ge \omega_2$. If, on the other hand, Martin's axiom and $\omega_1 < 2^\omega$ hold and B is an infinite cBA satisfying ccc, the algebra $C = RO(X^\omega)$ satisfies ccc and $|C| \le (\omega \cdot |B|)^\omega = |B|$.

<u>4.4</u>. Let κ be an infinite cardinal. A subset D of a cBA B is said to be κ-closed if, for any descènding sequence $(d_\alpha)_{\alpha<\lambda}$ in D where $\lambda<\kappa$ there is some $d \in D$ such that $d \le d_\alpha$ for every $\alpha < \lambda$. If B has a dense κ-closed subset, then B is (λ,∞)-distributive for every $\lambda < \kappa$. It is proved in [12] that, if B has a dense κ-closed subset D then B

may be completely embedded into a rigid cBA C which has a dense κ-closed subset. Moreover, if κ is regular and $\mu = \kappa \cdot |D|$ then $c(C) \le (\mu^{\underset{\smile}{\kappa}})^+$. We give an analogue of this result and 4.2:

<u>4.5. Theorem</u>. Let B be a cBA, κ a regular cardinal and D a dense κ-closed subset of B. Then B may be completely embedded into a homogeneous cBA C with a dense κ-closed subset E such that, if $c(B) \le \alpha^+$ for some α and $\mu = \alpha \cdot \kappa$, we have $c(C) \le (2^\mu)^+$.

<u>Proof</u>. We may assume $1_B \in D$. B is isomorphic to RO(D) where D has as a basis the sets $\{d \in D | d \le e\}$ for $e \in D$. Let λ be an infinite cardinal such that $\kappa \le \lambda$. Let

$$E = \{x \in D^\lambda | \ |\{i \in \lambda | x_i \ne 1_B\}| < \kappa\}$$

be endowed with the coordinate-wise partial order induced by D^λ. So E is a κ-closed partial order. Let $C = RO(E)$; we may assume that E is a dense subset of C. $c(C) \le (2^\mu)^+$ follows by 3.13 in [2]. The proof that C is homogeneous is completely analoguous to that given in 4.1.

<u>4.6</u>. We turn to the question about the possible cardinals of Aut B where B is a cBA and begin with a few trivial remarks. Firstly, if $b, c \in B$ and $b \le c$, then $|\text{Aut } B \upharpoonright b| \le |\text{Aut } B \upharpoonright c|$: for $f \in \text{Aut } B \upharpoonright b$, let $\bar{f} \in \text{Aut } B \upharpoonright c$ be the extension of f to $B \upharpoonright c$ satisfying $\bar{f}(x) = x$ for $x \le c \mathbin{\cdot\!-} b$. Clearly, the function mapping f to $\bar{f}$ is one-one.

If $B \cong \prod_{i \in I} B_i$ where $(B_i)_{i \in I}$ is a family of pairwise totally different cBA's then $\text{Aut } B \cong \prod_{i \in I} \text{Aut } B_i$ (see [7]).

If C is a rigid cBA and $2 \le n < \omega$, $|\text{Aut}(C^n)| = |C|$ follows from the representation of $\text{Aut}(C^n)$ given in [7].

If B has no rigid factor, then $|B| \le |\text{Aut } B|$: for $b \in B$, choose, by 3.3, b_1 and $b_2 \in B$ satisfying $b = b_1 \dot{+} b_2$ and $\tau(b_1) = \tau(b_2)$. Let $f_b \in \text{Aut } B$ such that f_b interchanges b_1 and b_2 and $f_b(x) = x$ for $x \le -b$.

The function mapping b to f_b is one-one since

$$b = -\sum\{x \in B \mid f_b(y) = y \text{ for every } y \leq x\}.$$

An arbitrary cBA B has, up to isomorphism, the form

$$B = A \times \prod_{i \in I} H_i^{\nu_i} \times \prod_{j \in J} E_j^{\mu_j} \times D$$
$$= A \times H \times E \times D$$

where A is atomic, i.e. $A \cong P(\alpha)$ for some cardinal α; $(H_i)_{i \in I}$ is a family of pairwise totally different atomless homogeneous cBA's; $(E_j)_{j \in J}$ is a family of pairwise totally different atomless rigid cBA's; ν_i and μ_j are cardinals such that $0 < \nu_i, \mu_j$ for $i \in I$, $j \in J$; D is an atomless cBA without homogeneous or rigid factors.

<u>4.7. Theorem E</u>. Let B be a cBA. Then Aut B is finite iff, in the decomposition of 4.6, α is finite, $|H \times D| = 1$ and $\mu_j = 1$ or $|E_j| = 1$ for $j \in J$ (i.e. E is rigid); hence $|\text{Aut } B| = \alpha!$. If Aut B is infinite , $|\text{Aut } B|^{\omega} = |\text{Aut } B|$.

<u>Proof</u>. We have

(*) $$\text{Aut } B \cong \text{Aut } A \times \prod_{i \in I} \text{Aut}(H_i^{\nu_i}) \times \prod_{j \in J} \text{Aut}(E_j^{\mu_j}) \times \text{Aut}(D).$$

If α is finite and $B \cong P(\alpha) \times E$ where E is rigid, Aut B is, up to isomorphism, the group of all permutations of α. Conversely, suppose Aut B and hence Aut A, $\text{Aut}(H \times D)$ and, for every $j \in J$, $\text{Aut}(E_j^{\mu_j})$ are finite. So $\alpha < \omega$. If $2 \leq \mu_j$ and $1 < |E_j|$ for some $j \in J$,

$$\omega \leq |E_j| = |\text{Aut}(E_j^2)| \leq |\text{Aut}(E_j^{\mu_j})|$$

by 4.6 and since E_j is infinite. $H \times D$ is an atomless cBA without rigid factors. So, if $|H \times D| > 1$, we get

$$\omega \leq |H \times D| \leq \text{Aut}(H \times D)$$

by 4.6.

Now let Aut B be infinite. By (*) it is sufficient to prove

$|\text{Aut } B|^{\omega} = |\text{Aut } B|$ for the cases $B = A$, $B = H_i^{\nu_i}$ and $B = E \times D$. If $B = H_i^{\nu_i}$ for some $i \in I$, $B^{\omega} = B$ which gives rise to an embedding of $(\text{Aut } B)^{\omega}$ into $\text{Aut}(B^{\omega}) \cong \text{Aut } B$. So let $B = E \times D$, i.e. B has no homogeneous factor. By 3.6, $R = R(B)$ is atomless. For $r \in R \setminus \{0\}$, let

$$\varphi(r) = |\text{Aut } B \restriction r|.$$

φ is a monotone function from $R \setminus \{0\}$ to the class of cardinals. Call $r \in R \setminus \{0\}$ φ-homogeneous if $\varphi(r') = \varphi(r)$ for every $r' \in R$ such that $0 < r' \leq r$. So

$$X = \{r \in R \setminus \{0\} \mid r \text{ is } \varphi\text{-homogeneous}\}$$

is a dense subset of R. Let $P \subseteq X$ be a partition of R. Since R is atomless, we may assume that, if $\varphi(p) = \gamma$ for some $p \in P$, $\{p \in P \mid \varphi(p) = \gamma\}$ is infinite.

Now

$$\text{Aut } B \cong \prod_{p \in P} \text{Aut}(B \restriction p)$$

since $B \restriction p$ and $B \restriction p'$ are totally different for $p \neq p'$. So

$$(**) \qquad |\text{Aut } B| = \prod_{p \in P} \varphi(p) = |\text{Aut } B|^{\omega}$$

since every factor $\varphi(p)$ occurring in (**) occurs infinitely often.

4.8. Let, in the decomposition of B given in 4.6,

$$E_1 = \prod \{E_j^{\mu_j} \mid \mu_j = 1\}, \quad E_2 = \prod \{E_j^{\mu_j} \mid \mu_j \geq 2\}.$$

It may be proved, by considering the cases $B = A$, $B = H \times D$ and $B = E$ that $|B| \leq |\text{Aut } B|$ if $|B| = |A \times H \times E_2 \times D|$ and $|\text{Aut } B| \geq \omega$.

References

[1] B. Balcar, P. Štěpánek: Embedding theorems for Boolean algebras and consistency results an ordinal definable sets. The Journal of Symbolic Logic 42 (1977), 64-75.

[2] W.W. Comfort, S. Negrepontis: The Theory of Ultrafilters. Berlin-Heidelberg-New York 1974.

[3] E. van Douwen, J.D. Monk, M. Rubin: Some questions about Boolean algebras, preprint by the Forschungsinstitut für Mathematik, ETH Zürich.

[4] G. Grätzer: General Lattice Theory. Basel-Stuttgart 1978.

[5] S. Grigorieff: Intermediate submodels and generic extensions in set theory. Annals of Mathematics 101 (1975), 447-490.

[6] A. MacIntyre: Model completeness for sheaves of structures. Fundamenta Mathematicae 81 (1973), 73-89.

[7] R. McKenzie, J.D. Monk: On automorphism groups of Boolean algebras. Colloquia mathematica societatis Janos Bolyai, 10. Infinite and Finite Sets, 951-988.

[8] H. Rasiowa: An Algebraic Approach to Non-Classical Logics. Amsterdam 1974.

[9] M. Rubin: On the reconstruction of Boolean algebras from their automorphism groups, to appear in: Archiv für mathematische Logik.

[10] R. Sikorski: Boolean algebras. Berlin-Heidelberg-New York 1964.

[11] R. Solovay, S. Tennenbaum: Iterated Cohen extensions and Souslin's Problem. Annals of Mathematics 94 (1970), 201-245

[12] P. Štěpánek: Cardinal collapsing and ordinal definability, preprint.

[13] A. Tarski: Cardinal Algebras. New York 1949.

[14] D.A. Vladimirov: Boolesche Algebren. Berlin 1972.

PSEUDO REAL CLOSED FIELDS

Alexander Prestel*)
Fakultät für Mathematik
Universität Konstanz
D-7750 Konstanz

In his famous paper [Ax] on the elementary theory of finite fields Ax considered fields K with the property that every absolutely irreducible affine variety defined over K has a K-rational point. These fields have been later called pseudo algebraically closed (pac) by Frey [Fr] and extensively studied by Jarden [Jr] and Wheeler [Wh]. The class of pac-fields turned out to be very interesting.

The above definition of pac-fields can be put into the following equivalent version: for every extension field L of K , if K is algebraically closed in L , then K is existentially closed in L . It has been this characterization of pac-fields which Basarab generalized in [Ba] for ordered fields. Basarab calls an ordered field (K,<) pseudo real closed (prc) if (K,<) is existentially closed in every ordered extension (L,<) in which it is algebraically closed.+)

---

*) The original intention of the author has been to publish in this volume a straight-forward proof of the elimination of quantifiers for the elementary theory of a special class of prc-fields. During his stay in Brasil in summer 1980 the author obtained various results on prc-fields. Therefore he decided to give here a comprehensive introduction to the theory of prc-fields.

+) We have been informed by S. Basrab that this notion of pseudo real closed fields has also been introduced by K.Mc Kenna in his thesis, indepently of [Ba].

One of the most natural questions to ask is whether every ordered algebraic extension $(K',<')$ of a prc-field $(K,<)$ is a prc-field too, a property which holds for pac-fields ([Ax], § 14, Lemma 1). Unfortunately the answer is no. By standard arguments (compare Theorem (1.1)) a prc-field $(K,<)$ can be constructed such that $\mathbb{Q}$ is algebraically closed in K. Hence $K' = K(\sqrt{2})$ admits at least 2 orderings. On the other hand a prc-field is uniquely ordered (in [B] this has been stated as a problem): Let $a \in K$ be positive. The absolutely irreducible polynomial $X^2 + Y^2 - a$ defines an absolutely irreducible plane curve over K. The ordering $<$ of K extends to the function field L of this curve over K (again denoted by $<$). Then $(K,<)$ is algebraically closed in $(L,<)$. Obviously, the existential sentence:

$$(\exists xy)\ x^2 + y^2 - a = 0$$

holds in $(L,<)$, hence also in $(K,<)$. This shows that a is a sum of two squares in K.

One of the main results (Theorem (3.1)) of this paper is that every algebraic extension K' of a prc-field is again a prc-field, if one extends the definition of prc suitably: Let K be a formally real field. Then we call K <u>pseudo</u> <u>real</u> <u>closed</u> (prc) if K is existentially closed in every extension field L to which all orderings of K extend, and in which K is algebraically closed. The crucial point here is that we do not fix one (or a certain number) of orderings on K. In case K admits exactly one ordering we come back to Basarab's definition. This is clear from the fact that the uniquely determined ordering $<$ in this case is definable by an existential sentence:

$$0 \leq a \quad \text{iff} \quad (\exists xy)\ a = x^2 + y^2$$

The Extension Theorem above also holds for non-real fields K'. In this case K' turns out to be a pac-field. Thus we may drop the condition of K being formally real. Now pac-fields (of characteristic zero) are special cases of prc-fields which seems to be natural in this approach.

The Extension Theorem will be proved in Section 3 . In Section 1 we investigate several algebraic properties of prc-fields. In Section 2 we deal with the model theory of prc-fields which have exactly n orderings. We also show in Section 2 (Theorem (2.1)) that a field K having only a finite number of orderings is a prc-field if and only if every absolutely irreducible plane curve which has a simple point in every real closure of K has an infinity of K-rational points. This 'curve condition' is already sufficient to guarantee that all orderings of K induce different topologies on K and that K is dense in all its real closures (Propositions (1.4) and (1.6)). In section 4 we prove that the class of prc-fields is elementary.

In this paper we make free use of standard model theoretic notions and results. We refer the reader to the book [Ch-K] of Chang-Keisler or the excellent thesis [v.d.D] of van den Dries.

0. Preliminaries

In this section we give some definitions and collect some algebraic results which will be frequently used. All fields considered in this paper will have characteristic zero.

For a field K we denote by $\tilde{K}$ its algebraic closure. Since we deal only with fields of characteristic zero, we may call a field extension L/K regular if K is (relatively) algebraically closed in L . Two extension fields L_1, L_2 of a common subfield K are called linearly disjoint over K if the tensor product $L_1 \otimes_K L_2$ is an integral domain. By Theorem 2, Ch. III of [La] we have

(0.1) L/K *is a regular field extension if and only if* $L \otimes_K \tilde{K}$ *is a field*.

From Theorem 4, Ch.III in [La] and Corollary 1, p.203 in [Ja] we get

(0.2) *Let* L_1/K *be regular and* L_2/K *be arbitrary. Then* L_1 *and* L_2 *are linearly disjoint over* K *and* $\mathrm{Quot}(L_1 \otimes_K L_2)$ *is a regular extension of* L_2 .

We call a subset V of some $\tilde{K}^m$ a K-variety if V is the set of zeros of some prime ideal $I \subset K[X_1,\ldots,X_m]$. By its *function field over* K we mean $F(V) = \mathrm{Quot}(K[\vec{X}]/I)$. If it happens that I is *absolutely prime*, i.e. the ideal generated by I in $\tilde{K}[\vec{X}]$ is still prime, then V is called *absolutely irreducible*. In this case K is a field of definition for V (see [La], p. 70). Hence we also say that V is an *absolutely irreducible variety defined over* K . From the discussion in [La] we get

(0.3) *Let* V *be a* K-*variety. Then* V *is absolutely irreducible if and only if* $F(V)/K$ *is a regular extension*.

By an *ordering* P of K we mean a subset P of K such that $P+P \subset P$, $P \cdot P \subset P$, $P \cap -P = \emptyset$, and $P \cup -P = K \setminus \{0\}$. Actually, P is the positive cone of an ordering defined by:

$$a <_P b \text{ iff } b - a \in P$$

By $(\overline{K,P})$ we denote the real closure of K with respect to P . It is uniquely determined up to isomorphism in $\tilde{K}$. The following fact is well-known; for convenience we give a proof of it.

(0.4) *Let* V *be an absolutely irreducible variety defined over* K , $F(V)$ *its function field over* K , *and* $f_1,\ldots,f_s \in K[\vec{X}]$. *Then an ordering* P *of* K *extends to* $F(V)$ *such that (the residue classes of)* $f_1,\ldots,f_s$ *are positive, if and only if* V *has a simple* $(\overline{K,P})$-*rational point* $\vec{a}$ *with* $f_1(\vec{a}),\ldots,f_s(\vec{a})$ *positive*.

Proof: First assume that there exists such an extension Q of P to $F(V)$. Let the prime ideal I of V be generated by $g_1,\dots,g_r \in K[X_1,\dots X_m]$ Then the following statement is satisfied in $(F(V),Q)$ by the residue classes of $X_1,\dots,X_m$ modulo I :

$$(\exists \vec{x})\left[\bigwedge_{j=1}^{s} f_j(\vec{x}) \in Q \wedge \bigwedge_{i=1}^{r} g_i(\vec{x}) = 0 \wedge \operatorname{rank}\left(\frac{\partial g_i}{\partial x_j}\right) = m - d\right]$$

where d is the dimension of V, i.e. the transcendence degree of $F(V)$ over K. Note that this statement can be actually expressed by an existential sentence in the language of ordered fields. Hence by the model completeness of real closed fields it also holds in $(\overline{K,P})$.

Conversely, let $\vec{a} \in (\overline{K,P})$ be a simple point of V such that $f_j(\vec{a})$ is positive. The local ring A of $\vec{a}$ in the function field $\bar{F}(V)$ of V over $(\overline{K,P})$ is regular. If M denotes its maximal ideal, then A/M is 'equal' to $(\overline{K,P})$ and $f_1 + M,\dots,f_s + M$ are positive in A/M. Since A is regular, its completion $\hat{A}$ is isomorphic to some power series ring R over $(\overline{K,P})$. By well-known methods one extends the ordering of $(\overline{K,P})$ to an ordering on R which then induces an ordering Q on $\mathrm{Quot}(A) = \bar{F}(V)$ such that $f_1,\dots,f_s \in Q$.

q.e.d.

Note that we only used the fact that V remains irreducible over $(\overline{K,P})$. The next assertion will be useful in Section 2. It is essentially van den Dries' Lemma (2.5), Ch. II in [v.d.D]. Since his proof is model theoretic, we give an algebraic one for convenience.

(0.5) Let L_1/K and L_2/K be regular extensions. Then to every pair P_1, P_2 of orderings on L_1, L_2 resp., inducing the same ordering P on K, there is a common extension Q to $L_1 \otimes_K L_2$.

Proof: Consider the set

$$Q' = \left\{ \sum_{i=1}^{N} (p_i \otimes q_i) x_i^2 \mid 1 \leq N \in \mathbb{N},\ p_i \in P_1,\ q_i \in P_2,\ x_i \in R \right\} \setminus \{0\},$$

where we let $R = L_1 \otimes_K L_2$. This set is a preordering on R , i.e. we have $Q' + Q' \subset Q'$, $Q' \cdot Q' \subset Q'$, $Q' \cap -Q' = \emptyset$, and $x^2 \in Q'$ for all $x \in R$. We have to prove only the third assertion. It suffices to show that

$$\sum_{i=1}^{N} (p_i \otimes q_i) x_i^2 = 0$$

can only hold in case all x_i vanish. Since this sum involves only a finite number of elements we may assume that L_1 and L_2 are finitely generated over K . Hence we may consider them as function fields of certain absolutely irreducible varieties V_1 and V_2 defined over K. Moreover, we may assume that p_i and q_i are regular functions. Since P_1 and P_2 both extend P , by (0.4), there are $\vec{a}, \vec{b} \in (\overline{K,P})$ such that $\vec{a}$ is a simple point of V_1 with $p_i(\vec{a})$ positive and $\vec{b}$ is a simple point of V_2 with $q_i(\vec{b})$ positive (for $1 \leq i \leq N$). Now $\vec{c} = (\vec{a}, \vec{b})$ is a simple point of the product variety $V = V_1 \times V_2$ such that $(p_i \otimes q_i)(\vec{c})$ is positive for all $1 \leq i \leq N$. Thus by (0.4) there is an extension of P to $F(V) = L_1 \otimes_K L_2$ such that all $p_i \otimes q_i$ are positive. Hence $\Sigma(p_i \otimes q_i)x_i^2 = 0$ implies $x_i = 0$ for all $1 \leq i \leq N$.

By standard arguments (cf.[Pr]) we can extend the preordering Q' to an ordering Q of the integral domain $L_1 \otimes_K L_2$. Obviously, Q extends P_1 as well as P_2 . q.e.d.

We now let $X_K = \{P \mid P \text{ ordering of } K\}$. There is a topology on X_K , the Harrison topology, generated by the (clopen) sets $H(a) = \{P \in X_K \mid a \in P\}$ where $a \in K \setminus \{0\}$. Endowed with this topology X_K becomes a totally disconnected compact space, the <u>order space</u> of K (compare [M-H], Lemma (2.8), Ch. III or [Pd], Theorem (6.5)). A field K is called formally real or just <u>real</u> if $X_K \neq \emptyset$. A field extension L/K is called <u>totally real</u> if the restriction map from X_L to X_K is surjective, i.e. if each $P \in X_K$ extends to some ordering on L .

1. Pseudo real closed fields

In this section we prove the existence of prc-fields and investigate algebraic properties of such fields. Recall that all fields are supposed to have characteristic zero.

A field K is called pseudo real closed (prc) if K is existentially closed in every totally real regular extension L of K.

Existentially closed here means that every existential sentence φ of the type

$$(\exists x_1,\dots,x_m)\ [p_1(\vec{x}) = 0 \wedge \dots \wedge p_r(\vec{x}) = 0 \wedge q(\vec{x}) \neq 0]$$

which holds in L also holds in K. The sentence φ is a formula in the language of fields with parameters from K, i.e. $p_1,\dots,p_r$ are polynomials with coefficients from K. As we will see later (Theorem (1.7)) in case X_K is finite this already implies that K is existentially closed in L even in the language augmented by predicates for each $P \in X_K$.

First we observe that by [Wh], Theorem 2.2, every pac-field K (of characteristic zero) is also a prc-field. In this case we have $X_K = \emptyset$.

Among the real fields every real closed field K is prc. This follows from the model completeness of the theory of real closed fields: Let $\bar{L}$ be the real closure of L with respect to some extension of the unique ordering of K to L. Then an existential sentence φ with parameters from K which holds in L also holds in $\bar{L}$ and hence in K. To obtain further examples of prc-fields let us fix some field k together with a system $(P_i)_{i\in I}$ of orderings of k. Consider the following class of ordered fields:

$M = \{(K,(P_i)_{i\in I}) \mid K/k$ is regular, P_i extends $p_i\}$

It is clear that M is closed under unions of (linearly ordered) chains. Thus there are elements $(K,(P_i)_{i\in I})$ of M which are M-existentially closed, i.e. if some $(K',(P_i')_{i\in I}) \in M$ extends $(K,(P_i)_{i\in I})$ then every existential sentence (even in the language augmented by predicates for each P_i) with parameters from K which holds in $(K',(P_i')_{i\in I})$ also holds in $(K,(P_i)_{i\in I})$ (compare [Ch], Ch. III, § 1). We claim hat K is prc. To prove this let L be a totally real regular extension of K. Choose an extension Q_i of the ordering P_i to L. Now obviously $(L,(Q_i)_{i\in I})$ belongs to M. Hence every existential sentence with parameters from K which holds in L also holds in K. We thus proved

(1.1) THEOREM. <u>To each field</u> k <u>and each system</u> $(p_i)_{i\in I}$ <u>of orderings on</u> k <u>there is a regular extension</u> K <u>of</u> k <u>such that</u> K <u>is prc and every ordering</u> p_i <u>extends to</u> K.

We will see later (the remark after Proposition (1.6)) that, in $(K,(P_i)_{i\in I})$ above, X_K in general need not coincide with $\{P_i \mid P_i$ extends $p_i\}$. In case X_K is finite, equality will hold (Proposition (1.6)).

If we let $k = \mathbb{Q}$ in the above construction and p be the unique ordering of $\mathbb{Q}$, we obtain a prc-field K which will be uniquely ordered but does not have $\sqrt{2}$ as an element. Hence K is not real closed.

The next theorem gives an algebraic characterization of prc-fields. We could have equally well taken this as definition.

(1.2) THEOREM. K <u>is a prc-field if and only if every absolutely irreducible variety</u> V <u>defined over</u> K <u>which has a simple</u> $(\overline{K,P})$-<u>rational point for each</u> $P \in X_K$ <u>has a K-rational point.</u>

<u>Proof</u>: Let K be prc and let the prime ideal of V be generated by some $g_1,\dots,g_r \in K[X_1,\dots,X_m]$. Then the existential sentence given by

$$(\exists\vec{x})\ [g_1(\vec{x}) = 0 \wedge\dots\wedge g_r(\vec{x}) = 0]$$

obviously holds in the function field F(V) of V over K . Since V is absolutely irreducible, F(V)/K is regular by (0.3). By (0.4) we know that each $P \in X_K$ extends to F(V) . Thus φ holds in K .

Conversely, let L be a totally real regular extension. Assume that the existential sentence ψ given by

$$(\exists x)\ [p_1(\vec{x}) = 0 \wedge\dots\wedge p_r(\vec{x}) = 0 \wedge q(\vec{x}) \neq 0]$$

with $p_1,\dots,p_r$, $q \in K[\vec{x}]$ holds in L . Then also the sentence

$$(\exists\vec{x}y)\ [p_1(\vec{x}) = 0 \wedge\dots\wedge p_r(\vec{x}) = 0 \wedge yq(\vec{x}) - 1 = 0]$$

holds in L . Fix some of those $\vec{x}$, $y \in L$. Let $F = K(\vec{x},y)$. Obviously F is the function field over K of the K-variety V defined by the prime ideal $I = \{f \in K[\vec{X},Y] \mid f(\vec{x},y) = 0\}$. Since F/K is regular, by (0.3) we know that V is absolutely irreducible. By (0.4) and the fact that L/K is totally real we know that V has a simple $(\overline{K,P})$-rational point for each $P \in X_K$. Hence, by assumption, there is a K-rational point $(\vec{a},b)$ on V. Thus in particular we have

$$p_1(\vec{a}) = 0 \wedge\dots\wedge p_r(\vec{a}) = 0 \wedge bq(\vec{a}) - 1 = 0$$

This shows that φ also holds in K .

q.e.d.

In the following, we will apply this criterion only to plane curves. For convenience we introduce the 'curve condition'.

C_K^Z : For every absolutely irreducible $f(X,Y) \in K[X,Y]$ which has a simple zero in $(\overline{K,P})$ for each $P \in Z$, there are $x,y \in K$ with $f(x,y) = 0$.

Here Z is a fixed subset of X_K. If $Z = X_K$, we just write C_K instead of C_K^Z. Note that every prc-field K satisfies C_K by Theorem (1.2).

(1.3) PROPOSITION. Let K satisfy C_K^Z. For all $a,b \in K \setminus \{0\}$ there is $c \in K$ such that for each $P \in X_K$ we have $c \in P$ iff $a,b \in P$, i.e. $H(a) \cap H(b) = H(c)$.

Proof*): Consider the plane curve defined by the polynomial

$$f(X,Y) = abX^2Y^2 + aX^2 + bY^2 - 1 = aX^2(bY^2+1) + (bY^2-1)$$

Easy inspection of f as polynomial in X proves irreducibility over $\tilde{K}[Y]$. Clearly, f has simple zeros in $(\overline{K,P})$ for all $P \in X_K$. Now let (x,y) be a K-rational point of this curve, i.e. $f(x,y) = 0$. The element

$$c := ab(ax^2 + by^2) = ab - a^2b^2x^2y^2$$

then satisfies the assertion of the proposition. Note that $c = 0$ is not possible. q.e.d.

Using the language of order spaces, Proposition (1.3) tells us that C_K^Z implies K to be a SAP-field (see [Pr], Proposition (6.6))

---

*) This proof is due to J. Kr. Arason, whom the author would like to thank for many helpful discussions on this subject.

(1.4) PROPOSITION. Let K satisfy C_K^Z. Then K is dense in $(\overline{K,P})$ for all $P \in X_K$.

Proof: Given $P \in X_K$ we have to prove that $(\overline{K,P})$ is contained in $(\widehat{K,P})$, the completion of K with respect to the uniformity induced by P. It suffices to show that every irreducible monic $g \in K[X]$ which has a zero $(\overline{K,P})$ also has a zero in $(\widehat{K,P})$. If g has no zero in $(\widehat{K,P})$, then the set $g(K)$ does not approach 0 (see [Ka], Theorem 1). Hence also $g^2(K)$ does not. Thus there is $\varepsilon \in P$ such that $0 <_P g^2(x) - \varepsilon^2$ for all $x \in K$. Now by Elimination of Quantifiers for real closed fields there is a formula without quantifiers involving only the coefficients $\vec{a}$ of g and a predicate Π for the positive cone (ordering) which holds in (K,Q) if and only if g has a zero in $(\overline{K,Q})$, where $Q \in X_K$ is arbitrary. Taking this formula in disjunctive normal form the fact that g has a zero in $(\overline{K,P})$ implies that some disjunctive part, say

$$p(\vec{a}) = 0 \wedge q_1(\vec{a}) \in \pi \wedge \ldots \wedge q_r(\vec{a}) \in \Pi ,$$

holds in (K,P) where $p, q_1, \ldots, q_r \in \mathbb{Z}[\vec{X}]$. Iterating Proposition (1.3) we find $c \in K$ such that for each $Q \in X_K$:

$$c \in Q \quad \text{iff} \quad q_1(\vec{a}) \in Q \wedge \ldots \wedge q_r(\vec{a}) \in Q .$$

Hence for all $Q \in X_K$ such that $c \in Q$ we know that g has a zero in $(\overline{K,Q})$. We may assume that $1 <_P c$. Now consider the polynomial

$$f(X,Y) = Y^2 + cg^2(X) - \varepsilon^2 \in K[X,Y] .$$

Inspection over $\tilde{K}[X]$ proves that f is absolutely irreducible. If $c \in Q$, there is some zero $\alpha \in (\overline{K,Q})$ of g. Thus (α, ε) is a simple zero of f in $(\overline{K,Q})$. If $c \in -Q$, we have the simple zero $(0, \sqrt{\varepsilon^2 - cg^2(0)})$ in $(\overline{K,Q})$. Hence by C_K^Z there is a K-rational point (x,y) of f. But this implies

$$g^2(x) = c^{-1} \cdot (\varepsilon^2 - y^2) \leq_P \varepsilon^2$$

a contradiction. q.e.d.

(1.5) PROPOSITION. Let K satisfy C_K^Z . Then every element of $\bigcap_{P \in Z} P$ is a sum of two squares in K .

Proof: Let $a \in \bigcap_{P \in Z} P$. Consider the absolutely irreducible polynomial

$$f(X,Y) = Y^2 + X^2 - a \in K[X,Y]$$

Obviously $(\sqrt{a}, 0)$ is a simple zero of f in $(\overline{K,P})$ for each $P \in Z$. Hence there are $x,y \in K$ such that $y^2 + x^2 = a$. q.e.d.

(1.6.) PROPOSITION: Let K satisfy C_K^Z . Then the orderings $P \in X_K$ induce different topologies on K . If Z is finite we get $Z = X_K$.

Proof: Assume $P_1, P_2 \in X_K$ induce the same topology on K . Then $\widehat{(K,P_1)} = \widehat{(K,P_2)}$ since the completion depends only on the topology. By Proposition (1.4) K is dense in $(\overline{K,P_1})$. Hence $\widehat{(K,P_1)}$ is also the completion of $(\overline{K,P_1})$. Therefore it is real closed. Let $\hat{P}$ be the unique ordering of $\widehat{(K,P_1)}$. Then we get $P_1 = K \cap \hat{P} = P_2$.

Now let Z be finite and assume there is some $P \in X_K \setminus Z$. Since P and the elements Q of Z induce pairwise different topologies on K , by the Approximation Theorem for V-Topologies (see[Pr-Z Theorem (4.1)), there is $a \in K$ which is P-close to -1 and Q-close to 1 for all $Q \in Z$. But then by Proposition (1.5), $a \in \bigcap_{Q \in Z} Q$. Thus a is a sum of two square and cannot be P-close to -1 . q.e.d.

Remark: If we drop finiteness of Z , by the proof of (1.6) we obtain that there is no $a \in K$ such that $H(a) \cap Z = \emptyset$. Since for SAP-field the sets $H(a)$ form a base of the Harrison topology on X_F, this tells us that C_K^Z implies density of Z in X_K. Actually, this is the best we can get: Let us consider the example $k = \mathbb{Q}(X)$ where $z = \{p \mid p$ non-archimedean ordering of $k\}$ is a dense proper subset of X_k (compare [Pr], Theorem (9.9)). Choose the system $(p_i)_{i\in I}$ such that $z = \{p_i \mid i\in I\}$. By Theorem (1.1) and its proof we get a field K which satisfies C_K^Z with $Z = \{P_i \mid i\in I\}$. Since the restriction map from X_K to X_k is continuous and X_K is compact, the density of z in X_k implies $Z \neq X_K$.

(1.7) THEOREM. Let X_K be finite, say $X_K = \{P_1,\dots,P_n\}$. Then K is a prc-field if and only if $(K,P_1,\dots,P_n)$ is existentially closed in every extension $(L,Q_1,\dots,Q_n)$ such that L/K is regular.

Note that 'existentially closed' is understood here in the language of fields augmented by predicates $\Pi_1,\dots,\Pi_n$ for the n fixed orderings.

Proof: Assume that K is prc and let $(L,Q_1,\dots,Q_n)$ be an extension of $(K,P_1,\dots,P_n)$ such that L/K is regular. By Theorem (1.1) and its proof there is an extension $(K',P_1',\dots,P_n')$ of $(L,Q_1,\dots,Q_n)$ such that K'/L is regular and K' satisfies $C_{K'}^{\{P_1',\dots,P_n'\}}$. Consider an existential sentence φ of the type

$$(\exists\vec{x})\ [p(\vec{x}) = 0 \wedge \bigwedge_{i=1}^{r} q_i(\vec{x}) \in \Pi_{\nu_i}]$$

where $p,q_i \in K[\vec{X}]$. Assume φ holds in $(L,Q_1,\dots,Q_n)$. Then it also holds in $(K',P_1',\dots,P_n')$. We choose some $\vec{a} \in K'$ such that $p(\vec{a}) = 0$ and $q_i(\vec{a}) \in P_{\nu_i}'$ for all $1 \le i \le r$. By Proposition (1.6), $P_1,\dots,P_n$ induce different topologies on K. Hence by the Approximation Theorem (see proof of (1.6)) for each $1 \le i \le r$ we find $c_i \in K$ which has the same sign with respect to P_j as $q_i(\vec{a})$ for all $1 \le j \le n$. Thus

$c_iq_i(\vec{a})$ is a sum of two squares in K' by Proposition (1.5). Now the existential sentence ψ (in the language of fields with parameters from K)

$$(\exists\vec{x})\left[p(\vec{x}) = 0 \wedge \bigwedge_{i=1}^{r} (\exists u_i v_i)\ c_i q_i(\vec{x}) = u_i^2 + v_i^2\right]$$

holds in K'. Since K'/K is regular and K is a prc-field, ψ also holds in K. But then φ clearly holds in $(K,P_1,\ldots,P_n)$.

q.e.d.

2. The elementary theory of prc-fields with n orderings

In this section we consider only prc-fields having exactly n orderings. Since n may be 0 too, the case of pac-fields (of characteristic zero) is included.

The structures under consideration now are of the type

$$(K,\ P_1,\ldots,P_n)$$

where K is a field and $P_1,\ldots,P_n \in X_K$. Such a structure will be called an <u>n-ordered field.</u>

First observe that the curve condition $C_K^{\{P_1,\ldots,P_n\}}$ can be expressed by a recursive set of elementary sentences in the language of n-ordered fields. What we will actually use in this section is the following slightly stronger 'curve condition'

(C): <u>For every absolutely irreducible</u> $f(X,Y) \in K[X,Y]$, <u>monic in</u> Y <u>such that there are</u> $x_i,y_i \in \overline{(K,P_i)}$ <u>with</u> $f(x_i,y_i) = 0$ <u>and</u> $\frac{\partial f}{\partial Y}(x_i,y_i) \neq 0$ <u>for all</u> $1 \leq i \leq n$, <u>and every</u> $h(X) \in K[X] \setminus \{0\}$ <u>there are</u> $a,b \in K$ <u>with</u> $f(a,b) = 0$ <u>and</u> $h(a) \neq 0$.

The class of n-ordered fields satisfying the condition (C) is an elementary class since for fixed (total) degree of f and h we may express (C) by a formula in the language of fields, augmented by predicates $\Pi_1,\dots,\Pi_n$ which stand for the n orderings. This is clear since "f is irreducible over $\tilde{K}$" can be already expressed in K by some quantifier free formula and "there are $x_i, y_i \in (\overline{K,P_i})$ with $f(x_i,y_i) = 0$ and $\frac{\partial f}{\partial Y}(x_i,y_i) \neq 0$ " can be expressed in (K,P_i) by some quantifier free formula for each $1 \leq i \leq n$.

The next theorem in particular shows that the class of n-ordered fields $(K,P_1,\dots,P_n)$ such that K is prc and $X_n = \{P_1,\dots,P_n\}$ is an elementary class. The theory of this class will be called the <u>theory of n-ordered prc-fields</u>. We denote it by PRC_n . It is axiomatized by the recursive set of formulas expressing the curve condition (C).

(2.1) THEOREM. <u>Let</u> $(K, P_1,\dots,P_n)$ <u>be an n-ordered field. Then</u> $(K, P_1,\dots,P_n)$ <u>satisfies (C) if and only if</u> K <u>is a prc-field and</u> $X_K = \{P_1,\dots,P_n\}$.

<u>Proof</u>: First let K be a prc-field with $X_K = \{P_1,\dots,P_n\}$. Assume f and h satisfy the assumptions of (C) . Then the existential sentence

$$(\exists xy)\,[f(x,y) = 0 \wedge h(x) \neq 0]$$

holds in K, because it obviously holds in the function field $L = \mathrm{Quot}(K[X,Y]/(f))$ of the variety defined by f over K , and L/K is a totally real regular extension.

Next, assume that (C) holds for $(K,P_1,\dots,P_n)$. Since (C) implies $C_K^{\{X_1,\dots,X_n\}}$, we know from Proposition (1.6) that $X_K = \{P_1,\dots,P_n\}$. Now let L be a totally real regular extension of K . Assume that a certain existential sentence φ (in the language of fields) with parameters from K holds in L . Then φ already holds in some subfield of L , finitely generated over K . Hence we may assume that L itself is finitely generated over K . We are going to embed L (as a field) into some elementary extension $(K^*, P_1^*,\dots,P_n^*)$ of $(K,P_1,\dots,P_n)$ which we assume to be $|K|^+$-saturated. Then φ will hold in K^* and hence in K .

Since L/K is regular, it is possible to find generators $x_1,\dots,x_m,x,y$ of L over K such that

(i) $x_1,\dots,x_m,x$ are algebraically independent over K

(ii) y is algebraic over $K(x_1,\dots,x_m,x)$ and there is $f \in K(x_1,\dots,x_m)[X,Y]$ monic in Y , such that $f(x,y) = 0$ und f is irreducible over $\widetilde{K(x_1,\dots,x_m)}$.

(For a proof of this fact see [Sch], Theorem 3D, Ch. V). We would now like to embed first $K(\vec{x}) = K(x_1,\dots,x_m)$ into K^* and then apply the condition (C) to the image of the absolutely irreducible polynomial $f(X,Y)$ in order to embed finally L . However, in order to verify the assumptions on a simple zero in each real closure $(\overline{K^*,P_i^*})$, we have to take care of the orderings too.

Hence let $Q_1,\dots,Q_n$ be extensions of $P_1,\dots,P_n$ to L resp. We are now going to embed $(K(\vec{x}),Q_1',\dots,Q_n')$ into $(K^*,P_1^*,\dots,P_n^*)$ where $Q_i' = Q_i \cap K(\vec{x})$. Consider the diagram of the structure $(K(\vec{x}),Q_1',\dots,Q_n')$. Since $(K^*,P_1^*,\dots,P_n^*)$ is $|K|^+$- saturated we only have to realize there every finite subset of this diagram, which then is a formula φ of the type

$$p(\vec{x}) \neq 0 \wedge \bigwedge_{j=1}^{r} q_j(\vec{x}) \in \Pi_{\nu_j}$$

There is no non-trivial equality to take care of, since $x_1,\dots,x_n$ are algebraically independent. Separating by the index of Π and replacing $p(\vec{x}) \neq 0$ by $p^2(\vec{x}) \in \Pi_1$ we may assume that

$$\varphi = \varphi^{(1)} \wedge \ldots \wedge \varphi^{(n)}$$

where each $\varphi^{(i)}$ is of the type

$$p_1^{(i)}(\vec{x}) \in \Pi_i \wedge \ldots \wedge p_{m_i}^{(i)}(\vec{x}) \in \Pi_i$$

Since $(\exists\vec{x})\varphi^{(i)}$ holds in $(\overline{L,Q_i})$ it also holds in $(\overline{K^*,P_i^*})$ by Elimination of Quantifiers for real closed fields. Since $(K^*,P_1^*,\ldots,P_n^*)$ satisfies (C), by Proposition (1.4), K^* is dense in $(\overline{K^*,P_i^*})$. Now $(\exists\vec{x})\varphi^{(i)}$ also holds in (K^*,P_i^*), since $\varphi^{(i)}$ defines an open set in $(\overline{K^*,P_i^*})$. Thus for each $1 \leq i \leq n$ the P_i^*- open set defined by $\varphi^{(i)}$ is non-empty. By Proposition (1.6) and the Approximation Theorem for V-Topologies (see [Pr-Z], Theorem (4.1)) we may conclude that the intersection of these sets is non-empty too. Hence φ is realizable in $(K^*,P_1^*,\ldots,P_n^*)$.

Now we can identify $(K(\vec{x}),Q_1',\ldots,Q_n')$ with some substructure of $(K^*,P_1^*,\ldots,P_n^*)$. Finally we will apply (C) in $(K^*,P_1^*,\ldots,P_n^*)$ to the absolutely irreducible polynomial $f(X,Y)$. This is possible since the validity of

$$(\exists xy)\ [f(x,y) = 0 \wedge \frac{\partial f}{\partial Y}(x,y) \neq 0]$$

now carries over from $(\overline{L,Q_i})$ to $(\overline{K^*,P_i^*})$ for each $1 \leq i \leq n$. By (C), for all $a_1,\ldots,a_N \in K^*$ we find $a,b \in K^*$ such that $f(a,b) = 0$ and $h(a) = \prod_{j=1}^{N}(a-a_j) \neq 0$. Using again $|K|^+$- saturatedness of K^* we may therefore assume that a is transcendental over $K(\vec{x})$. But then $K(\vec{x},a,b)$ is isomorphic to L.

q.e.d.

It should be mentioned that the original geometric idea behind the above proof is the following: consider L as the function field of some absolutely irreducible variety V. Find a generic curve on V

and spezialize it suitably. The difficulty in doing so is that the spezialized curve should still have a simple point in each real closure of K . This difficulty is reflected in the above proof by the fact that the embedding of $K(\vec{x})$ into K^* had to respect the orderings.

The next Lemma will imply that the existential theorems of PRC_n form a decidable set.

(2.2) **LEMMA.** Let $(K',P_1',\dots,P_n')$ and $(K'',P_1'',\dots,P_n'')$ be two models of PRC_n . Assume there is an n-ordered common substructure $(K,P_1,\dots,P_n)$ such that K'/K and K''/K are regular field extensions. Then an existential sentence with parameters from K holds in $(K',P_1',\dots,P_n')$ if and only if it holds in $(K'',P_1'',\dots,$

Proof: Consider the tensor product $K' \otimes_K K''$. Since K'/K and K''/K are regular, by (0.2) we know that $K' \otimes_K K''$ is an integral domain and that $L = \mathrm{Quot}(K' \otimes_K K'')$ is a regular extension of K' and K'' . By (0.5) we find orderings $Q_1,\dots,Q_n$ on L which are common extensions of $P_1',\dots,P_n'$ and $P_1'',\dots,P_n''$ resp.

Now every existential sentence φ with parameters from K which holds in $(K',P_1',\dots,P_n')$ trivially holds in $(L,Q_1,\dots,Q_n)$. By Theorem (1.7), φ then also holds in $(K'',P_1'',\dots,P_n'')$. Similarly we argue for the way back.

q.e.d.

(2.3) **THEOREM.** The subset of PRC_n consisting of boolean combinations of existential sentences is decidable.

Proof: Follow essentially the proof of Theorem (2.1), Ch. II in [v.d.D] . Replace $\overline{OD_n}$ by PRC_n , and use Lemma (2.2), Theorem (1.1) and the fact that it is decidable whether $f \in \mathbb{Q}[X]$ has a zero in an effectively given algebraic number field.

q.e.d.

Now we consider among the models $(K_1,P_1,\ldots,P_n)$ of PRC_n such ones which do not have any algebraic extension $(L,Q_1,\ldots,Q_n)$. This again is an elementary class since we may express the maximality of $(K,P_1,\ldots,P_n)$ by

(A): Every monic irreducible $f \in K[X]$ which has a zero in (K,P_i) for each $1 \le i \le n$ is linear.

By the use of Elimination of Quantifiers for real closed fields one sees that (A) can be expressed by a recursive set of elementary sentences, The theory axiomatized by (A) and (C) will be denoted by $\overline{PRC_n}$. As we will see in the next theorem $\overline{PRC_n}$ coincides with van den Dries' theory $\overline{OD_n}$. This provides a seemingly weaker axiomatization of $\overline{OD_n}$ (compare [v.d.D], Theorem (1.2), Ch. II).

Note that $\overline{PRC_o}$-models are algebraically closed fields and $\overline{PRC_1}$-models are real closed fields.

(2.4) THEOREM. $\overline{PRC_n} = \overline{OD_n}$

Proof: $\overline{OD_n}$ is the unique model companion of OF_n, the theory of n-ordered fields. Hence it suffices to show that $\overline{PRC_n}$ is also a model companion of OF_n.

Given an n-ordered field $(k,p_1,\ldots,p_n)$ there are M-existentially closed extensions $(K,P_1,\ldots,P_n)$ in the class

$$M = \{(K,P_1,\ldots,P_n) \mid (k,p_1,\ldots,p_n) \text{ substructure of } (K,P_1,\ldots,P_n)\}$$

As in the proof of Theorem (1.1) we obtain that $(K,P_1,\ldots,P_n)$ satisfies (C) and hence is a model of PRC_n. Assume (A) does not hold in $(K,P_1,\ldots,P_n)$. Then there is a finite algebraic extension $(L,Q_1,\ldots,Q_n)$ of $(K,P_1,\ldots,P_n)$ in which for a certain irreducible non-linear polynomial $f \in K[X]$ the existential sentence $(\exists x)\ f(x) = 0$ holds. Since $(L,Q_1,\ldots,Q_n) \in M$ it also must hold in $(K,P_1,\ldots,P_n)$, a contradiction.

It remains to show that $\overline{PRC_n}$ is model complete. Let $(K,P_1,\dots,P_n) \subset (L,Q_1,\dots,Q_n)$ be two models of $\overline{PRC_n}$. By (A), L is a regular extension of K. Now Theorem (1.7) implies that every existential sentence with parameters from K which holds in $(L,Q_1,\dots,Q_n)$ also holds in $(K,P_1,\dots,P_n)$. This proves model completeness of PRC_n.

q.e.d.

In [v.d.D] van den Dries shows that there is an extension $\widetilde{OD}_n$ by definitions of $\overline{OD_n}$ which admits Elimination of Quantifiers. The new predicates he adds are less natural then those used by Kiefe [K] for the theory of pseudo-finite fields. In Example (2.16) van den Dries showes that Kiefe's predicates do not suffice to obtain Elimination of Quantifiers. We will now show that adding just n new individual constants makes Kiefe's predicates sufficient.

Let $\widetilde{PRC}_n$ be the elementary theory in the language of n-ordered fields, augmented by constants $c_1,\dots,c_n$ and (m+1)-ary predicates W_m for each $m \geq 1$, axiomatized by (A), (C), $-c_i \in \Pi_i \wedge \bigwedge_{j \neq i} c_i \in \Pi_j$ for each $1 \leq i \leq n$, and

$$(\forall x_o,\dots,x_m)\,[\,W_m(x_o,\dots,x_m) \longleftrightarrow (\exists y)\, x_o + x_1 y + \dots + x_m y^m = 0\,]$$

(2.5) THEOREM. $\widetilde{PRC}_n$ <u>admits Elimination of Quantifiers</u>.

<u>Proof</u>: By a well-known criterion for quantifier elimination we have to consider the following situation: Let

$$\underline{K}^{(i)} = (K^{(i)},\, P_1^{(i)},\dots,P_n^{(i)},\, c_1^{(i)},\dots,c_n^{(i)},\, (W_m^{(i)})_{m\geq 1})$$

be models of $\widetilde{PRC}_n$ for $i = 1,2$. Furthermore let $\underline{R} = (R,\dots)$ be a common substructure of $\underline{K}^{(1)}$ and $\underline{K}^{(2)}$ and let φ be a simple existential sentence (in the augmented language) with parameters from R. If then φ holds in $\underline{K}^{(1)}$, we have to show that it also holds in $\underline{K}^{(2)}$.

Passing to the field of quotients we may assume that R is a field. Since $W_m^{(1)}$ and $W_m^{(2)}$ coincide on R, the same polynomials $f \in R[X]$ which have a zero in $K^{(1)}$ also have a zero in $K^{(2)}$. It is well-known that this forces the algebraic closures of R in $K^{(1)}$ and $K^{(2)}$ to be isomorphic. If we can assume that such an isomorphism σ respects the orderings, we may assume that R is algebraically closed in $K^{(1)}$ and $K^{(2)}$. In fact, we can assure this: the constants $c_j^{(1)}$ and $c_j^{(2)}$ belong to R and coincide. Hence we just write c_j. Now consider some $a \in K^{(1)}$. Let c be the product of those c_j's which are negative whenever a is negative. Then ac is totally positive and hence by (A) is a square in $K^{(1)}$. Thus, if a is algebraic over R, $\sigma(ac)$ is a square in $K^{(2)}$. Hence a and $\sigma(a)$ have the same sign behaviour.

Now, since $K^{(1)}/R$ and $K^{(2)}/R$ are regular, we are in the situation of Lemma (2.2). We may assume that $\varphi \equiv (\exists x)\psi$ where ψ is the conjunction of negated and unnegated atomic formulas. If we could replace equivalently those atomic formulas involving some predicate W_m by some existential formula not involving any W_m, we could deduce from Lemma (2,2) that φ holds in $\underline{R}^{(2)}$.

To eliminate an atomic formula

$$W_m(a_o(x),\ldots,a_m(x))$$

just replace it by

$$(\exists y)\ a_o(x) + a_1(x)y + \ldots + a_m(x)y^m = 0$$

To eliminate all the negated formulas

$$\neg W_m(a_o(x),\ldots,a_m(x))$$

we proceed as follows: First we fix some $x \in K^{(1)}$ which satisfies ψ.

Now $\neg W_m(a_o(x),\ldots,a_m(x))$ means that the polynomial $f(Y) = a_o(x) + \ldots + a_m(x)Y^m \in K^{(1)}[Y]$ does not have any zero in $K^{(1)}$. We split f into irreducible factors over $K^{(1)}$, say $f = f_1 \cdot \ldots \cdot f_r$. By (A), each of the f_i's must be without zeros in $\overline{(K^{(1)},P^{(1)}_{\nu_i})}$ for some ν_i. By Elimination of Quantifiers for real closed fields, this can be expressed in $(K^{(1)},P^{(1)}_{\nu_i})$ by some quantifier free formula. Hence we see that we may replace $\neg W_m(a_o(x),\ldots,a_m(x))$ by an existential formula expressing: "there are (coefficients of) polynomial $f_1,\ldots,f_r$ such that f_i does not have a zero in $\overline{(K^{(1)},P^{(1)}_{\nu_i})}$ and $f = f_1 \cdot \ldots \cdot f_r$".

q.e.d.

3. Algebraic extensions

In this section we prove that every algebraic extension of a prc-field is again prc. We also show that in case k is a countable real hilbertian field there are many real algebraic extensions of k which are prc but not real closed. For an arbitrary real field k such extension need not exist (see Remark at the end of this section).

(3.1) EXTENSION THEOREM. Let K be a prc-field. Then every algebraic extension L of K is a prc-field too.

Proof: We first treat the case of a finite extension, say $[L:K] = n$. Let V be an absolutely irreducible variety defined over L which has a simple $\overline{(L,Q)}$-rational point for each $Q \in X_L$. Using Weil's descent method (see [We], § 1.3) we will construct an absolutely irreducible variety W defined over K which has a simple $\overline{(K,P)}$-rational point for each $P \in X_K$, and a morphism σ from W to V defined over L. Then, by the assumption on K, W will have a K-rational point whose image under σ is an L-rational point of V.

Let the prime ideal of V be generated by $g_1,\dots,g_r \in L[X_1,\dots,X_m]$. By $\rho_i(x) = x^{(i)}$ with $1 \leq i \leq n$ we denote the n different embeddings of L over K into $\tilde{K}$, say $\rho_1 = \mathrm{id}$. Then $g_1^{(i)},\dots,g_r^{(i)}$ generate the prime ideal of an absolutely irreducible variety $V^{(i)}$ defined over $L^{(i)}$. If we take new variables for each i, say $X_1^{(i)},\dots X_m^{(i)}$ and $g_1^{(i)},\dots,g_r^{(i)} \in L^{(i)}[X_1^{(i)},\dots,X_n^{(i)}]$, then $g_1^{(1)},\dots,g_r^{(1)},\dots$ $\dots,g_1^{(n)},\dots,g_r^{(n)}$ generate the prime ideal I of the absolutely irreducible product variety $V^{(1)} \times\dots\times V^{(n)}$. Let $\alpha_1,\dots,\alpha_n$ be a base of L over K. Then $\det(\alpha_j^{(i)}) \neq 0$ and hence

$$(*) \qquad X_l^{(i)} = \sum_{j=1}^{n} \alpha_j^{(i)} Y_{lj}$$

defines an invertible linear substitution for each $1 \leq l \leq m$. Call J the prime ideal which results from this substitution. J thus defines an absolutely irreducible variety W. In order to see that W is already defined over K, take as generators for J the uniquely determined polynomials f_{sj} which satisfy the linear system

$$\alpha_1^{(1)} f_{s1} + \dots + \alpha_n^{(1)} f_{sn} = g_s^{(1)}(X/Y)$$

$$\alpha_1^{(n)} f_{s1} + \dots + \alpha_n^{(n)} f_{sn} = g_s^{(i)}(X/Y)$$

where (X/Y) indicates the substitution $(*)$. Now it is easily seen that $f_{sj} \in K[\vec{Y}]$.

It remains to show that W has a simple $(\overline{K,P})$-rational point for each $P \in X_K$. Fix some $P \in X_K$ and write $\overline{K}$ for $(\overline{K,P})$. The embeddings ρ_i which map L into $\overline{K}$ correspond to the extensions Q of P to L. Say $\rho_1,\dots,\rho_r$ are these 'real' embeddings. By assumption, V has a simple $(\overline{L,Q})$-rational point for the corresponding Q's. This means that there is a simple point $(x_1^{(i)},\dots,x_m^{(i)}) \in \overline{K}$ in $V^{(i)}$ for each $1 \leq i \leq r$. The remaining 'non-real' embeddings

$\rho_{r+1},\dots,\rho_n$ map L into $\overline{K}(\sqrt{-1}) = \tilde{K}$. They come in pairs which are conjugate with respect to $\overline{K}$. In this case we choose an arbitrary simple point $(x_1^{(i)},\dots,x_m^{(i)}) \in \overline{K}(\sqrt{-1})$ for $V^{(i)}$ and the conjugate point $(\bar{x}_1^{(i)},\dots,\bar{x}_m^{(i)})$ for $V^{(j)}$ if ρ_j is conjugate to ρ_i over $\overline{K}$. (Here $\bar{x}$ means conjugation over $\overline{K}$). It is now clear that the corresponding point $(y_{11},\dots,y_{mn})$ on W is simple and has its coordinates in $\overline{K}$. Since K is a prc-field there is a K-rational point on W . The first m coordinates of its preimage under the substitution (*) form an L-rational point of V . This finishes the proof for finite extensions L/K .

The proof for an arbitrary algebraic extension L/K makes use of Theorem (4.1), proved in the next section. In Theorem (4.1) we will show that the property of L being prc can be expressed by a set of axioms in the first order language of fields. As will be seen from the proof of Theorem (4.1) these axioms can be chosen to be of the type $\varphi \equiv \forall \vec{u}\, \exists \vec{v}\ \ \psi(\vec{u},\vec{v})$ where ψ is a quantifier free formula. We have to show that all these axioms φ hold in L. Let $\vec{u}$ be a finite sequence of elements of L . Then $K(\vec{u})$ is a finite extension of K Hence $K(\vec{u})$ is prc by the above proof. Thus we find $\vec{v}$ in $K(\vec{u})$ such that $\psi(\vec{u},\vec{v})$ holds in $K(\vec{u})$. Hence it also holds in L .

q.e.d.

The only algebraic PRC_1-extensions of a given (real) field k we have seen so far are real closures of k . In [v.d.D] , Ch. II, Theorem (3.1), van den Dries proves the existence of $\overline{PRC_n}$ - models $(L,Q_1,\dots,Q_n)$ which are algebraic extensions of a countable hilbertian field k which admits n orderings inducing different topologies on k. But for $n = 1$ such an L still is real closed. Modifying slightly van den Dries' proof we will obtain in the next theorem algebraic PRC_1-extensions of k which are not real closed. Concerning information about hilbertian fields we refer the reader to [Ro].

(3.2) THEOREM. Let k be a countable real hilbertian field and (R,Q) a real closure of k. To every finite number of elements $\alpha_1,\ldots,\alpha_r \in R \setminus k$ there exists a PRC_1-subfield (K,P) of (R,Q) which omits $\alpha_1,\ldots,\alpha_r$.

Proof: We enumerate all pairs (f,h) such that $h \in R[X] \setminus \{0\}$, and $f \in R[X,Y]$ is absolutely irreducible, monic in Y, such that there are $a,b \in R$ with $f(a,b) = 0$ and $\frac{\partial f}{\partial Y}(a,b) \neq 0$. The enumeration is chosen such that every pair occurs infinitely often. We are going to construct a countable chain

$$k = K_o \subset K_1 \subset \ldots \subset K_m \subset \ldots \subset R$$

of finite extensions K_m of k. Thus every K_m is a hilbertian field as well (see [Ro]). The field we are looking for will be $K = \bigcup_{m \in \mathbb{N}} K_m$ together with its (unique) ordering $P = K \cap Q$.

Assume that we already constructed K_m. Then we consider the m-th pair (f,h). If either h or f do not have its coefficients in K_m, we take $K_{m+1} = K_m$. If both have coefficients in K_m, we will construct $K_{m+1} \supset K_m$ such that there are $a,b \in K_{m+1}$ with $f(a,b) = 0$ and $h(a) \neq 0$. To take care also of the elements $\alpha_1,\ldots,\alpha_r$ let $g = \prod_{j=1}^{r} \mathrm{Irr}(\alpha_i,k) \in k[Z]$ and assume g has no zero in k_m. We will then ensure that g has no zero in K_{m+1} too.

Now let $h \in K_m[X]$ and $f \in K_m[X,Y]$. By the assumption on f and the Implicit Function Theorem applied to (R,Q) we know that there is a Q-open set U such that for each $a \in U$, $f(a,Y)$ has a zero in (R,Q). By enlarging K_m if necessary, we may assume that the intersection $U \cap K_m$ is non-empty. From Theorem (1.16), Ch. II in [v.d.D] and its proof we know that there is an elementary extension (K_m^*, P^*) of $(K_m, K_m \cap Q)$ and an element $t \in K_m^* \setminus K_m$ such that $K_m(t)$ is algebraically closed in K_m^* and $f(t,Y)$ has a zero in $\overline{(K_m^*, P^*)}$. Let $F = K_m(t)[Y]/(f(t,Y))$. From the following diagram we conclude

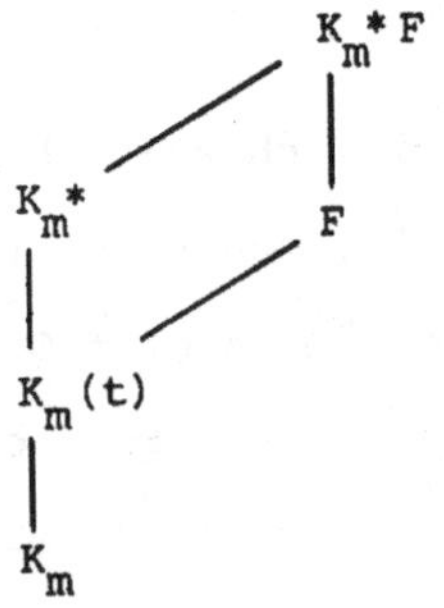

(i) $f(t,Y)$ is irreducible over K_m^*, and $h(t) \neq 0$

(ii) $f(t,Y)$ has a zero in $\overline{(K_m^*, P^*)}$

(iii) $g(Z)$ has no zero in $K_m^*[Y]/(f(t,Y))$.

(i) follows from the facts that t is transzendental over K_m and $K_m^*/K_m(t)$ is regular. (ii) follows from the choice of t. (iii) follows from the facts that

$$K_m{}^*F \simeq K_m{}^*[Y]/(f(t,Y))$$

and K_m is algebraically closed in $K_m{}^*F$. This last assertion is implied by the regularity of $K_m{}^*F/F$ (cf.(0.2)) and the regularity of F/K_m (cf.(0.3)).

It is clear how to express (i), (ii), and (iii) by an elementary formula φ in the language of ordered fields. Hence we know that the sentence $\exists t\varphi$ holds in (K_m^*, P^*) and thus also in $(K_m, K_m \cap Q)$. This means that we can find $a \in K_m$ such that $f(a,Y)$ is irreducible over K_m, $h(a) \neq 0$, and $f(a,Y)$ has a zero $b \in R$ such that g has no zero in $K_m(b) \simeq K_m[Y]/(f(a,Y))$. Now let $K_{m+1} = K_m(b)$

If we finally take $K = \bigcup_{m \in \mathbb{N}} K_m$, it becomes clear that $(K, K \cap Q)$ is a PRC_1-model such that $\alpha_1, \ldots, \alpha_r \notin K$.

q.e.d.

Remark: The assumption on k to be hilbertian is essential: Let k be henselian with respect to some valuation, and K/k a real algebraic prc-extension. Then by Proposition (1.4) and §8 in [Pr], the field K must be real closed.

4. The class of prc-fields is elementary

In this final section we prove just one theorem.

(4.1) THEOREM. The class of prc-fields can be axiomatized in the language of fields.

Proof: We will show that K is a prc-field if and only if for every $m \geq 0$ we have

(H_m): For every absolutely irreducible $f \in K[X_1, \ldots, X_{m+1}, Y]$, monic in Y, such that there are $\vec{x}, y \in (\overline{K,P})$ with $f(\vec{x}, y) = 0$ and $\frac{\partial f}{\partial Y}(\vec{x}, y) \neq 0$ for each $P \in X_K$, and every $h \in K[X_1, \ldots, X_{m+1}] \setminus \{0\}$, there are $\vec{a}, b \in K$ with $f(\vec{a}, b) = 0$ and $h(\vec{a}) \neq 0$.

It remains to express the collection of all (H_m) in the language of fields. The only problem is to eliminate for polynomials f of fixed total degree and arbitrary coefficients $\vec{c}$ the clause "for each $P \in X_K$ there are $\vec{x}, y \in (\overline{K,P})$ with $f(\vec{x},y) = 0$ and $\frac{\partial f}{\partial Y}(\vec{x},y) \neq 0$ ".

By Elimination of Quantifiers for real closed fields we can replace "there are ..." by a quantifier free formula involving P . If we take this formula in conjunctive normal form, we see that it suffices to express in the field language some clause of the type

"for each $P \in X_K : p(\vec{c}) = 0 \vee q_1(\vec{c}) \in P \vee \ldots \vee q_r(\vec{c}) \in P$ "

where $p, q_1, \ldots, q_r$ are polynomials with integral coefficients. So we have to formalize that there is no ordering on K such that all $q_i(\vec{c})$ are negative. This means that $-q_1(\vec{c}), \ldots, -q_r(\vec{c})$ cannot generate a preordering of K . To express this by a formula, we use the fact that in a prc-field every sum of 3 squares equals a sum of 2 squares by Proposition (1.5). We take this fact as a first axiom for prc-fields. Now we can formulate elementarily that there are sums s_i of squares, not all zero, such that $\sum_{i=1}^{N} b_i s_i = 0$, where the b_i range over all possible r+1-fold products of elements of $\{1, -q_1(\vec{c}), \ldots, -q_r(\vec{c})\}$.

It remains to prove that every field K satisfying all (H_m) is prc. As in the proof of Theorem (2.1) we have to embed a finitely generated, totally real regular extension $L = K(x_1, \ldots, x_{m+1}, y)$ into a $|K|^+$-saturated elementary extension K^* of K . Since we do not restrict to plane curves here, the situation is much easier.

Assume that $x_1, \ldots, x_{m+1}$ are algebraically independent over K and let $f \in K[X_1, \ldots, X_{m+1}, Y]$ be irreducible and monic in Y such that $f(\vec{x},y) = 0$. By (o.3), f is absolutely irreducible. Let $P^* \in X_{K^*}$ and $Q \in X_L$ such that $P^* \cap K = Q \cap K$. Then the validity of

$$(\exists \vec{x}, y)\ [f(\vec{x},y) = 0 \wedge \frac{\partial f}{\partial Y}(\vec{x},y) \neq 0]$$

carries over from $(\overline{L,Q})$ to $(\overline{K^*,P^*})$ by the Elimination of Quantifiers for real closed fields. By (H_m), there are $\vec{a},b \in K^*$ with $f(\vec{a},b) = 0$ and $h(\vec{a}) \neq 0$. In general, $\vec{a},b$ depend on the choice of h. However, since K^* is $|K|^+$-saturated, we even find $\vec{a},b \in K^*$ with $f(\vec{a},b) = 0$ and $h(\vec{a}) \neq 0$ for all $h \in K[\vec{X}] \setminus \{0\}$. Hence $a_1,\ldots,a_{m+1}$ are algebraically independent over K, and thus $K(\vec{a},b)$ is isomorphic to L.

q.e.d.

It should be mentioned that the restriction to hypersurfaces is not necessary to prove the axiomatizability of the class of prc-fields. But it makes life easier in writing down such axioms, and it is interesting in its own right. We do not know whether it is possible to restrict to plane curves, as we did in Theorem (2.1) in case of X_K being finite.

In a forthcoming paper we show that there is a certain elementary subclass of the class of prc-fields which is model complete. In fact, we will show that its theory is the model companion of pre-ordered fields. The existence of such a model companion has been conjectured by van den Dries (see [v.d.D], p. 73).

Last but not least, we would like to thank Peter Roquette for many inspiring discussions.

REFERENCES

[Ax] J. Ax: The elementary theory of finite fields. Ann. of Math. 88(1968), 239-271

[Ba] S.A. Basarab: Definite functions on algebraic varieties over ordered fields. To appear in Revue Roumaine Math. pures et appliquées.

[Ch] G. Cherlin: Model theoretic algebra; selected topics. Lecture Notes in Math. 521, Springer, 1976

[Ch-K] C. Chang-H. Keisler: Model theory. North-Holland, Amsterdam,1973

[v.d.D] L. van den Dries: Model theory of fields. Thesis, Utrecht, 1978

[Fr] G. Frey: Pseudo algebraically closed fields with non-archimedean real valuations. J. Alg. 26(1973), 202-207

[Ja] N. Jacobson: Lectures in abstract algebra III. van Nostrand, New Haven, 1964

[Jr] M. Jarden: Elementary statements over large algebraic fields. Trans. Amer. Math. Soc. 164(1972), 67-91

[Ka] I. Kaplansky: Polynomials in topological fields. Bull. Amer. Math. Soc. 54(1948), 909-916

[Ki] C. Kiefe: Sets definable over finite fields: their zeta-functions. Trans.Amer.Math.Soc. 223(1976), 45-59

[La] S. Lang: Introduction to algebraic geometry. Interscience, New York, 1958

[M-H] J. Milnor-D. Husemoller: Symmetric bilinear forms. Springer, Berlin-Heidelberg-New York, 1973

[Pr] A. Prestel: Lectures on formally real fields. Monografias de Matemática 22, IMPA, Rio de Janeiro, 1975

[Pr-Z] A. Prestel-M. Ziegler: Model theoretic methods in the theory of topological fields. J. reine angew. Math. 299/300 (1978), 318-341

[Ro] P. Roquette: Nonstandard aspects of Hilbert's irreducibility theorem, in: Model theory and algebra, 231-275, Lecture Notes in Math. 498, Springer, 1975

[Sch] W. Schmidt: Equations over finite fields. An elementary approach. Lecture Notes in Math. 536, Springer, 1976

[We] A. Weil: Adeles and algebraic groups. Institute for advanced studies, Princeton, 1961

[Wh] W. Wheeler: Model-complete theories of pseudo-algebraically closed fields. Ann. Math. Logic 17(1979), 205-226

Some Remarks on the Mathematical Incompleteness of Peano's Arithmetic Found by Paris and Harrington

T. von der Twer

1. Introduction

This paper is to supplement Paris' and Harrington's papers [1] and [2]. We shall assume the purely combinatorial lemmas of [1] (numbers 2.6 - 2.9, 2.12 - 2.14) on the strong partition relation $M \not\rightarrow (c)^d_e$ and give a simplified and more elaborate exposition of the main ideas of [1] and [2]. For most results, including the famous independence by Paris and Harrington, we can completely avoid provabilities in PA. Whenever these occur, they will be reduced to well-known ones explicitly. Finally, a couple of applications shall illustrate the model theoretic significance of the Σ_o-indiscernibles playing a central rôle in [1] and [2], e.g. on non-finite axiomatizability of PA, nonexistence of recursive models other than $\mathbb{N}$ of PRA, completions of PA satisfied by initial segments of models.

Basic Concepts and Notation

PA is first-order Peano's arithmetic, its language L containing $0,1,+,\cdot,<$. The Σ_o-formulas (or bounded formulas, or limited formulas) of L have only bounded quantifiers, e.g. $\exists x \leq y$, $\forall x < y$. Π_o is Σ_o . Σ_{n+1}-formulas, $n \in \mathbb{N}$, are of the shape $\exists x_1 \ldots \exists x_m \varphi$, where φ is a Π_n-formula. Dually, Π_{n+1}-formulas are the $\forall x_1 \ldots \forall x_m \varphi$ with $\varphi \in \Sigma_n$.

Models as well as their underlying sets are denoted by M, M_1, M' etc. . $\mathbb{N}$ is the natural model of PA, other models of the occurring theories are thought of as containing $\mathbb{N}$. $Th_{\Pi_n}(M)$ is the $Th(M) \cap \Pi_n$. We often write $M \models \varphi(a,b,c,\ldots)$ for formulas φ and $a,b,c,\ldots \in M$ instead of $M \models \varphi(x,y,z,\ldots)$ $[x/a, y/b, z/c,\ldots]$ or of $(M,a,b,c\ldots) \models \varphi(\underline{a},\underline{b},\underline{c},\ldots)$. For $n \in \mathbb{N}$ $\underline{n}$ is the natural constant term for n in L . Our notation widely does not distinguish between symbols s for arithmetically definable functions or relations and their meanings s^M in the model M , the difference being marked off by using $x,y,z,\ldots$ for L-variables and $a,b,c,\ldots$ for elements of models, in general.

For $a,b \in M$ $[a,b]$ will denote the interval of elements between a and b , including these. I is an initial segment of M iff $I \subseteq M$ and $\forall a,b \in M(a < b \in I \Rightarrow a \in I)$. An initial substructure I of M is an initial segment which is also a substructure. M is then an end-extension of I , which will be denoted by $I \subseteq_e M$. $I \subseteq_{\Sigma_n} M$ means that $I \subseteq M$ and for all Σ_n-formulas φ , $\vec{a}$ in I : $I \models \varphi(\vec{a})$ iff $M \models \varphi(\vec{a})$, i.e. I is a Σ_n-substructure of M . We have $I \subseteq_{\Sigma_0} M$ if $I \subseteq_e M$.

PRA (primitive recursive arithmetic) is arithmetic with additional symbols for any primitive recursive function and axioms for their defining equations plus induction axioms for the open formulas (with parameters) only, of the extended language. We shall use the primitive recursive truth-predicate tr_0 for Σ_0-formulas with a fixed sequence $x_1,\ldots,x_m$ of variables satisfying $PRA \vdash tr_0(\ulcorner\underline{\varphi}\urcorner, x_1,\ldots,x_m) \leftrightarrow$ $\leftrightarrow \varphi(x_1,\ldots,x_m), \varphi \in \Sigma_0$, our notation suppressing numeral- and substitution-functions, $\ulcorner\varphi\urcorner$ denoting the Gödel number of φ . For $n \geq 1$ we need Σ_n-definable truth-predicates tr_n satisfying $PA \vdash tr_n(\ulcorner\underline{\varphi}\urcorner, x_1,\ldots,x_m) \leftrightarrow \varphi(x_1,\ldots,x_m)$ for all $\varphi(x_1,\ldots,x_m) \in \Sigma_n$.

The Strong Partition Relation and the Combinatorial Principle α (Paris-Harrington Formula).

For $a,b,c,d,e \in \mathbb{N}$ $[a,b] \rightarrow (c)^d_e$ means: for all functions (called partitions) P from the set $[[a,b]]^d$ of d-element-subsets of $[a,b]$ into the set of numbers $<e$ there is a homogeneous set $X \subseteq [a,b]$ for P with $\mathrm{Card}(X) \geq c$. X is called homogeneous for P if $P(\vec{a})$ is the same for all $\vec{a} \in [X]^d$.

1.1 The Finite Ramsey Theorem reads:

$\forall a,c,d,e \in \mathbb{N}\ \ \exists b \in \mathbb{N}\ \ [a,b] \rightarrow (c)^d_e$.

This statement can be formalized in the language of PA, as follows: use Gödel's β-function to encode sequences arithmetically, regard the codes of strictly increasing sequences as sets and their lenghts as the cardinals of the respective sets, then build up the involved set-theoretical notions as usual. Note that any bounded arithmetically definable set in an $M \models PA$ can thus be encoded and hence be provided with a cardinal. The resulting Π_2-formalization of 1.1 is known to be a theorem of PA. Strenghtening the partition relation $[a,b] \rightarrow (c)^d_e$ by the condition that the required homogeneous sets X be relatively large, i.e. $\min(X) \leq \mathrm{Card}(X)$, one obtains a relation $[a,b] \twoheadrightarrow (c)^d_e$ which leads to the valid combinatorial principle

1.2 $\quad \forall a,c,d,e \in \mathbb{N}\ \ \exists b \in \mathbb{N}\ \ \ [a,b] \twoheadrightarrow (c)^d_e$.

The proof of this uses the Infinite Ramsey Theorem, cf. [1]. Like 1.1, 1.2 can be formalized as a Π_2-sentence, which was proved to be independent of $PA \cup Th_{\Pi_1}(\mathbb{N})$ by Paris and Harrington. Π_2 (or, by Matiyasevic's theorem, $\forall_2$) is the best attainable by their methods, whereas Matiyasevic's theorem leads to $\forall_1$-independences. (Recall that in the $\forall_n$, $\exists_n$-hierarchy also bounded quantifiers count.)

It is convenient to replace the formalization of 1.2 by

1.3 $\forall x \forall z \exists y \ [x,y] \underset{*}{\to} (z+1)^z_z$, henceforth denoted by α

which is equivalent in PA (by lemmas 2.7 and 2.8 of [1]) .
We shall use 1.3 , but remark that

1.4 $\forall z \exists y \ [0,y] \underset{*}{\to} (z+1)^z_z$

is again provably equivalent, by

$$PA \vdash ([0,y] \underset{*}{\to} (z+1)^z_{z\cdot\max(x+1,7)}) \to ([x,y] \underset{*}{\to} (z+1)^z_z$$

which is obtained by using lemmas 2.7 - 2.9, 2.12 in [1].

2. Basic Results on the Existence of Σ_0-Indiscernibles

The meaning of the easy combinatorial lemmas 2.6 - 2.9, 2.12 - 2.14 of [1] can be expressed in the single

2.1 Lemma: Let $r \in \mathbb{N}$, $g:\mathbb{N} \to \mathbb{N}$ a primitive recursive function. There is $d \in \mathbb{N}$, $d \geq r$, so that for all $M \models PA$ and $a,b,c \in M$, $c > d$, we have: If $M \models [a,b] \underset{*}{\to} (c)^d_d$, then for all partitions $P:[[a,b]]^r \to [0,2)$ which are encodable in M (cf. the remark following 1.1) there is a homogeneous set X , encoded in M , for P , $X \subset [a,b]$, such that

$M \models Card(X) \geq c \wedge \min(X) \geq d \wedge g(\min(X)) \leq Card(X) \wedge$
$\wedge \forall x \forall y \ (x \in X \wedge y \in X \wedge x < y \to g(x) < y)$.

Remark: For $M = \mathbb{N}$ 2.1 is completely explicitly proved in [1], the main steps are: (i) collecting together a fixed finite number of partitions into one, lifting superscripts d to a common one without lifting the subscripts e essentially, (ii) using the condition $Card(X) \geq \min(X)$ which distinguishes $\underset{*}{\to}$ from $\to$ to obtain the additional properties of X relative to a fixed primitive recursive g Let us mention an ugly misprint in [1] in lemma 2.14 . There it is important that s' does not depend on e . Now this proof is immediately realized to hold for arbitrary $M \models PA$, since all parameters but a,b,c may be kept standard finite; only little of PRA is needed.

The main idea behind the work of Paris and Harrington is the observation that $[a,b] \underset{*}{\to} (c)^d_e$ in some $M \models PA$ with c,d,e large enough leads to the existence of large enough subsets of [a,b] which are homogeneous for (sets of) Σ_o-formulas with parameters, i.e. sets of Σ_o-indiscernibles, and thus to establish initial segments of M which model PA again.

The next lemma 2.2 will give a technically simplified, though sharpened, version of what we need of the lemmas 2.10, 2.11 and 2.15 in [1], the idea of proof remaining the same.

2.2 <u>Lemma</u>: Let $s,t \in \mathbb{N}$, $t > 0, r = 2t+1$. Let g be the primitive recursive function given by $g(n) = n\cdot(2^{n\cdot s}+1)$. Let d satisfy the conditions of 2.1 for these r,g . Then we have for $M \models PA$,
$M \models [a,b] \underset{*}{\to} (c)^d_d$, $a,b,c \in M$, $c > d$: there is a sequence $(e_i | i \in M,\ M \models i < c)$, $e_i \in [a,b)$, $e_i^2 < e_{i+1}$, so that for all Σ_o-formulas $\varphi(x,y_1,\dots,y_t)$ with $\ulcorner\varphi\urcorner < s$ in the language of PA and all $i_o,i_1,\dots,i_t,\ j_1,\dots,j_t \in [0,c)$:
if $i_o < i_1 <\dots< i_t < j_1 <\dots< j_t$, then
$M \models \forall x < e_{i_o} (\varphi(x,e_{i_1},\dots,e_{i_t}) \leftrightarrow \varphi(x,e_{j_1},\dots,e_{j_t}))$.

<u>Proof</u>: With the assumptions, Lemma 2.1 can be applied to the partition of 2t+1-tuples in [a,b] ($\vec{v},\vec{w}$ denote t-tuples with $u < \vec{v} < \vec{w}$, $\vec{v} < \vec{w}$ means that any component of $\vec{v}$ is smaller than any of $\vec{w}$) defined by $P(u,\vec{v},\vec{w}) = 0$ iff ($M \models \forall x < u\ (\varphi(x,\vec{v}) \leftrightarrow \varphi(x,\vec{w}))$ for all Σ_o-formulas $\varphi(x,y_1,\dots,y_t)$ with $\varphi < s$), otherwise $P(u,\vec{v},\vec{w}) = 1$. So let X be a homogeneous set for P according to 2.1. We have $M \models \mathrm{Card}(X) \geq c$, so let e_i, $i < c$, be the first c elements of X .

By the choice of g $e_i^2 < e_{i+1}$. By $c > d \geq r = 2t+1$ there are 2t+1-tuples in $\{e_i | i < c\}$. Thus, to conclude the result on satisfaction of Σ_o-formulas below s it will suffice to prove that $P(u,\vec{v},\vec{w}) = 0$ for all 2t+1-tuples in X . Set $\bar{a} = \min(X)$. For $\varphi(x,y_1,\ldots,y_t) \in \Sigma_o$ with $\ulcorner\varphi\urcorner < s$, $z \in M$, $M \models z < \bar{a}$, $\vec{v}$ in $[X]^t$, set $P_{\varphi,z}(\vec{v})$ = truth-value of $\varphi(z,\vec{v})$ in M . There are at most $2^{\bar{a}\cdot s}$ sequences $(P_{\varphi,z}(\vec{v}) | \varphi,z)$, and by 2.1 and the choice of g we have $(\bar{a} \geq 2t+1)$: $c \geq g(\bar{a}) = \bar{a}\cdot(2^{\bar{a}\cdot s}+1) > t\cdot(2^{\bar{a}\cdot s}+1)$, so $[X \setminus \{\bar{a}\}]^t$ has a subset A of more than $2^{\bar{a}\cdot s}$ elements such that for $\vec{v},\vec{w} \in A$, $\vec{v} \neq \vec{w}$, we have $\vec{v} < \vec{w}$ or $\vec{w} < \vec{v}$. Now by the Dirichlet "pigeon-hole" principle (which is available by the induction principle) there are distinct $\vec{v},\vec{w} \in A$ with $(P_{\varphi,z}(\vec{v}) | \varphi,z) = (P_{\varphi,z}(\vec{w}) | \varphi,z)$, hence for these $P(\bar{a},\vec{v},\vec{w}) = 0$, and by homogeneity this holds for all 2t+1-tuples of X . Since Card(X) $\geq a$, the sequence of the e_i can be taken at least of lenght a .

<u>Remark</u>: Notice that the proof only deals with bounded definable subsets of M , so the steps familiar with the case $M = \mathbb{N}$ will remain possible for $M \models PA$.

The basic results on the existence of Σ_o-indiscernibles can now be established:

<u>2.3 Theorem</u>:

(i) For all finite sets Γ of Σ_o-formulas there is $n \in \mathbb{N}$ so that: If $M \models PA$, $M \models [a,b] \underset{*}{\rightarrow} (c)^n_n$, $c > n$, then there is a sequence $(e_i | i < \max(a,c))$, $a < e_i < b$, satisfying

(1) $e_i^2 < e_{i+1}$, $i+1 < \max(a,c)$,

(2) for all $\varphi(x,y_1,\ldots,y_t) \in \Gamma$, $i_o < i_1 <\ldots< i_t < j_1 <\ldots< j_t <$ $\max(a,c)$ $M \models \forall x < e_{i_o} (\varphi(x,e_{i_1},\ldots,e_{i_t}) \leftrightarrow \varphi(x,e_{j_1},\ldots,e_{j_t}))$.

(ii) (Paris) If $M \models PA$, $M \models [a,b] \underset{*}{\to} (c+1)^c_{c}$, $c > \mathbb{N}$, then there is a sequence $(e_i \mid i \in \mathbb{N})$, $e_i \in [a,b)$ satisfying

(1) $e_i^2 < e_{i+1}$, $i \in \mathbb{N}$

(2) for all $\varphi(x,y_1,\ldots,y_t) \in \Sigma_o$, $i_o < i_1 < \ldots < i_t, i_o < j_1 < \ldots < j_t$ in $\mathbb{N}$:

$M \models \forall x < e_{i_o} (\varphi(x,e_{i_1},\ldots,e_{i_t}) \leftrightarrow \varphi(x,e_{j_1},\ldots,e_{j_t}))$.

(3) Moreover, for $(e_i \mid i)$ as in (1),(2), $I = \{d \in M \mid \exists i \in \mathbb{N}\ d < e_i\}$ we have $I \subseteq_e M$ and $I \models PA$.

<u>Remark</u>: (ii) is stated in [2], with a slight modification, but not proved there. Note that up to 4.3 none of the results to follow will need any of 2.1 - 2.3 for arbitrary models of PA, but only for $\mathbb{N}$ itself or, sometimes, for $M \models Th(\mathbb{N})$.

<u>Proof</u>: (i) is immediate by 2.2, taking into account the final remark in the proof of 2.2 .

(ii) Let $\gamma(y)$ be a formula describing the primitive recursive set of the Gödel numbers of all the Σ_o-formulas

$\forall x < (z)_{i_o} (\varphi(x,(z)_{i_1},\ldots,(z)_{i_t} \leftrightarrow \varphi(x,(z)_{j_1},\ldots,(z)_{j_t}))$,

$i_o < i_1 < \ldots < i_t < j_1 < \ldots < j_t$ in $\mathbb{N}$, $(x)_y$ denoting $\beta(x,y+1)$, the y'th component of x (insert the Σ_o-arithmetic description of the graph of β). By (i), using lemma 2.8 in [1] to lift superscripts, we have for all $n \in \mathbb{N}$ $M \models \rho(n)$, where (tr_o is the truth-predicate for Σ_o-formulas)

$\rho(x) = \exists z\ \forall y < x((\gamma(y) \to tr_o(y,z)) \wedge (z)_y^2 < (z)_{y+1} \wedge \underline{a} \leq (z)_y < \underline{b})$.

The variable z might even be bounded by some nonstandard element of M. Since $M \models PA$, $\mathbb{N}$ is not definable with parameters in M, so for some $d \in M \setminus \mathbb{N}$ $M \models \rho(d)$. Choose $e \in M$ satisfying the existence formula $\rho(\underline{d})$, $((e)_i \mid i \in \mathbb{N})$ will do. For convenience, we give the proof of (3) in [1], which is not to be altered. Let e_i, I be as in (3), $d < e_{i_o}$, φ a Σ_o-formula in the language of PA. We have (i with subscript always denoting a natural number): $I \subseteq_e M$, by (1), and

$I \models \exists x_1 \forall x_2 \ldots \exists x_t \quad \varphi(d,x_1,\ldots,x_t)$

iff $\exists i_1 > i_o \forall i_2 > i_1 \ldots \exists i_t > i_{t-1}$: $I \models \exists x_1 < e_{i_1} \forall x_2 < e_{i_2} \ldots \varphi(d,\vec{x})$

iff $\exists i_1 > i_o \forall i_2 > i_1 \ldots \exists i_t > i_{t-1}$: $M \models \exists x_1 < e_{i_1} \forall x_2 < e_{i_2} \ldots \varphi(d,\vec{x})$

iff $M \models \exists x_1 < e_{i_o+1} \forall x_2 < e_{i_o+2} \ldots \exists x_t < e_{i_o+t} \varphi(d,\vec{x})$, by (2)

iff $I \models \exists x_1 < e_{i_o+1} \forall x_2 < e_{i_o+2} \ldots \exists x_t < e_{i_o+t} \varphi(d,\vec{x})$.

Thus, for any $i_o \in \mathbb{N}$ and arbitrary formula $\psi(x)$ there can be effectively found a bounded $\psi'(x)$ containing parameters among the e_i , so that $I \models \forall x < e_{i_o}(\psi(x) \leftrightarrow \psi'(x))$. By this, induction for I (without parameters) reduces to Σ_o-induction for I with parameters. The latter is satisfied, since it is in M and $I \subseteq_e M$. Now full induction without parameters implies the same with parameters, as is well-known. We might have avoided this by inserting sequences of "smaller" elements, instead of just one parameter, into the homogeneity-condition (2).

3. The Independence of the Strenghened Finite Ramsey Theorem and Some Applications of Its Proof.

The original proof [1] for $PA \nvdash \alpha$ in outline runs as follows: First a system T is introduced the models of which satisfy the basic axioms for $0,1,+,\cdot,<$ plus Σ_o-induction and contain a distinguished infinite set of Σ_o-indiscernibles as occur in 2.3,(ii). The main lemma in [1] is $PA \vdash \alpha \rightarrow Con(T)$ (for this the technical background exposed in paragraph 2 is needed), and by $PA \vdash Con(T) \rightarrow Con(PA)$ (cf. 2.3,(ii),(3)) and Gödel's Second Incompleteness Theorem saying $\nvdash Con(PA)$, $PA \nvdash \alpha$ can be concluded. Thus a lot of carrying out proofs in PA is required, not counting Gödel's theorem. Instead, we can directly attack the subsequent 3.3 (theorem 3.2 in [1]) which occurs in [1] as a corollary of the proof for $PA \nvdash \alpha$, and then we shall obtain the independence of α as an immediate corollary, the proof of 3.3 relying only on 2.2 for $M = \mathbb{N}$. Then we notice that this simplified proof for PA $\nvdash \alpha$ comes very close to the one given in [2] which in [2] is based on 2.3,(ii) and which is easily boiled down to an application of 2.3,(ii) for $M \models Th(\mathbb{N})$, thus again avoiding formal proofs within PA. As the second way still entails tr_o , contrary to the first one, it may be worth while to give both of them.

First we define the above-mentioned system T :

<u>3.1 Definition</u>: The language for the system T includes $0,1,+,\cdot,<$ and constants $c_i, i \in \mathbb{N}$. T contains the axioms:

(1) the usual axioms for first-order well-ordering theory and the recursive definitions of $+,\cdot$,

(2) induction axioms for all Σ_o-formulas (with parameters) in the language of PA,

(3) $c_i^2 < c_{i+1}$, $i \in \mathbb{N}$,

(4) for all Σ_o-formulas $\varphi(x,y_1,\ldots,y_t)$ in the language of PA , $i_o < i_1 < \ldots < i_t < j_1 < \ldots < j_t$ in $\mathbb{N}$, the axioms:
$\forall x < c_{i_o} (\varphi(x,c_{i_1},\ldots,c_{i_t}) \leftrightarrow \varphi(x,c_{j_1},\ldots,c_{j_t}))$.

<u>Remark</u>: We could safely replace here $\vec{x}$ by x (cf. the proof of 2.3, (ii), (3)) and the condition $i_o < i_1 < \ldots < i_t$, $i_o < j_1 < \ldots < j_t$ by the one in (4).

By means of 3.1 we can state in a handy form what is needed of 2.2:

<u>3.2 Lemma</u>: For all finite $S \subseteq T$ there is $d \in \mathbb{N}$ such that for all $a,b \in \mathbb{N}$: If $[a,b] \underset{*}{\to} (d+1)^d_d$, then there are $k \in \mathbb{N}$, $e_o,\ldots,e_k \in [a,b]$ with $(\mathbb{N}, e_o,\ldots,e_k) \models S$, e_i interpreting c_i .

Since $\mathbb{N} \models \alpha$, we have a total recursive function $f:\mathbb{N} \to \mathbb{N}$ given by $f(a) = \mu b([a,b] \underset{*}{\to} (a+1)^a_a)$. This function exceeds any provably total recursive function (provability in PA):

3.3 Theorem (Paris, Harrington): If $g:\mathbb{N} \to \mathbb{N}$ is a provably total recursive function, then $\exists n \in \mathbb{N} \quad \forall m \in \mathbb{N} \quad (m \geq n \Rightarrow f(m) > g(m))$.

Proof: Let $g:\mathbb{N} \to \mathbb{N}$ be a recursive function, thus its graph is Σ_1-definable, with $\forall n \in \mathbb{N} \quad \exists m \geq n \ (f(m) \leq g(m))$. We want to show that g is not provably total. Now let $S \subseteq T$ (the T of 3.1), S finite, d for S as in 3.2, $a \geq d$, $a \in \mathbb{N}$, $b = f(a) \leq g(a)$.
By 3.2 there are $e_o, \ldots, e_k \in [a,b]$ so that
$(\mathbb{N}, a, e_o, \ldots, e_k) \models S \cup \{c \leq c_o\} \cup \{\forall x < c_i(\neg g(c) \simeq x) \ i \leq k\}$,
a interpreting the new constant c, e_i the c_i, $i \leq k$, "$g(y) \simeq x$" abbreviating the arithmetical definition of the graph of g.
By compactness, the system
$T' = T \cup \{c \leq c_o\} \cup \{\forall x < c_i(\neg g(c) \simeq x) \mid i \in \mathbb{N}\}$ is consistent.
Take a model M' of this, then the $c_i^{M'}$ by 2.3,(ii),(3) determine an $I' \subseteq_e M'$, $I' \models PA$. If we had $I' \models g(c) \simeq d$, then we had (contradicting $M' \models T'$) $M' \models g(c) \simeq d \wedge d < c_i$ for some $i \in \mathbb{N}$, since $g(y) \simeq x$ is a Σ_1-formula and $I' \subseteq_{\Sigma_o} M'$. As we might have added $Th(\mathbb{N})$ to T', we obtain

3.4 Corollary (Paris, Harrington): $PA \cup Th_{\Pi_1}(\mathbb{N}) \nvdash \alpha$, noting that α just states the totality of f. By the remark to 1.4 the recursive function $a \to \mu b[0,b] \underset{*}{\to} (a+1)_a^a$ exceeds any provably total recursive function, too.

Second proof of 3.4: Let $M \models Th(\mathbb{N})$, $a \in M \setminus \mathbb{N}$. Since $N \models \alpha$, there is a smallest $b \in M$ so that $M \models [a,b] \underset{*}{\to} (a+1)_a^a$. Thus by 2.3(ii) (only for models of true arithmetic) there is an $I \subseteq_e M$, $I \models PA$, $a \in I < b$. If there were $b' \in I$ with $I \models [a,b'] \underset{*}{\to} (a+1)_a^a$, we had ($\Sigma_1$-formulas going up to end extensions) $M \models [a,b'] \underset{*}{\to} (a+1)_a^a$, which contradicts the minimality of b. Hence $I \models \neg\alpha$, and by $I \subseteq_e M \quad I \models Th_{\Pi_1}(\mathbb{N})$.

<u>Remark</u>: In [2] Paris defines an indicator for $M \models PA$ to be a definable function $Y: M \times M \to M$ such that for all $a,b \in M$: $Y(a,b) > \mathbb{N}$ iff there is an $I \subseteq_e M$, $I \models PA$, $a \in I < b$. He shows that if the graph of Y has a Σ_1-definition which defines an indicator in any countable $M \models PA$, then the sentence $\forall x \forall z \exists y \quad Y(x,y) > z$ is true but unprovable in PA, the latter requiring but the "only-if"-part of the indicator-property. We just stated in 2.3,(ii), that the function $Y(a,b) = \max\{c \mid [a,b] \underset{*}{\to} (c+1)^c_c\}$ satisfies this part of the indicator-property and exploited this for unprovability. Note that any indicator for all countable $M \models PA$ is an indicator for any $M \models PA$, which can easily be proved by a result of Gaifman's in [5] reading: If $M_1 \subseteq M_2$, $M_1, M_2 \models PA$, then with $M_3 = \{a \in M_2 \mid \exists b \in M_1\ a<b\}$ one has $M_1 \prec M_3 \subseteq_e M_2$. This also shows that in the definition of an indicator $I \subseteq_e M$ can be replaced by $I \subseteq M$ at all.

H. Friedman showed in [4] that if M_1, M_2 are countable models of PA, $Th_{\Sigma_1}(M_1) \subseteq Th(M_2)$, $SSy(M_1) = SSy(M_2)$, where $SSy(M_1)$, the standard system of M_1, is $\{X \subset \mathbb{N} \mid X = \{n \in \mathbb{N} \mid M \models \varphi(n)\}$ for some formula φ with parameters in $M\}$, <u>then</u> M_1 is isomorphic to an initial segment of M_2. Furthermore, if $M_1 \subseteq_e M_2$, $M_1 \neq \mathbb{N}$, $M_1, M_2 \models PA$, then $SSy(M_1) = SSy(M_2)$, cf. [3]. Taking this into account, α turns out to be a nice example for Theorem 2 in [3] which Ehrenfeucht and D. Jensen asked for:

<u>3.5 Corollary</u>: There are models $M_1, M_2 \models PA$ which are mutually initial segments (up to isomorphism) of one another such that $M_1 \models \alpha, M_2 \models \neg\alpha$.

<u>Proof:</u> Let T' be as in the proof of 3.3, $M_1' \models Th(\mathbb{N}) \cup T'$, M_1' countable. Define M_2' to be the initial substructure of M_1' determined by $\{c_i^{M_1'} \mid i \in \mathbb{N}\}$, M_1, M_2 the reducts of M_1', M_2' to the language of PA. Since $\mathbb{N} \subseteq_e M_2 \subseteq_e M_1$, $Th_{\Sigma_1}(M_1) = Th_{\Sigma_1}(M_2)$ and $SSy(M_1) = SSy(M_2)$, hence M_1 and M_2 are mutually initial segments,

and by $M_1' \models T'$, $M_2 \models \neg\alpha$.

As a second application, we can reprove Ryll-Nardzewski's theorem that PA is not finitely axiomatizable. First we give to $PA \nvdash \alpha$ the form of an ω-incompleteness:

<u>3.6 Lemma</u>: All sentences $\alpha_n = \forall x \exists y[x,y] \underset{*}{\rightarrow} (\underline{n}+1)^{\underline{n}}_{\underline{n}}$, $n \in \mathbb{N}$, are provable in PA.

<u>Proof</u>: The sentences $\exists y[0,y] \underset{*}{\rightarrow} (\underline{n}+1)^{\underline{n}}_{\underline{n}}$, being true Σ_1-sentences, are provable, now use the remark to 1.4.
Let us mention that even the $\forall x \forall y \exists z[x,y] \underset{*}{\rightarrow} (z)^{\underline{n}}_{y}$ are provable, but for this one needs a definable version of the Infinite Ramsey Theorem for fixed superscripts, using induction "from outside" on these. Cf. [6] for this topic. But here 3.6 will do.

2.3,(i) once being proved, the following theorem by Paris and Wilkie can be provided with a proof which is by far simpler than the game-theoretic one given in [2]. In fact, 3.7 is a corollary of 2.3,(i)

<u>3.7 Theorem (Paris, Wilkie)</u>: For all finite $S \subseteq PA$ there is $n \in \mathbb{N}$ so that $S \nvdash \alpha_n$, even $S \cup Th_{\pi_1}(\mathbb{N}) \nvdash \alpha_n$. In particular, PA is not finitely axiomatizable.

<u>Remark:</u> By using truth-predicates, the result can easily be extended to: PA with induction restricted to Σ_r-formulas for fixed $r \in \mathbb{N}$ does not imply all the α_n.

<u>Proof</u>: Let S be a finite subset of PA. By the proof of 2.3,(ii),(3) a finite set Γ of Σ_0-formulas $\varphi(x,y_1,\ldots,y_t)$ matches with the induction axioms contained in S (which may be assumed without parameters) in such a way that
<u>if</u> $(M,(e_i)_{i \in \mathbb{N}}) \models PA \cup \{c_i^2 < c_{i+1} \mid i \in \mathbb{N}\} \cup T_\Gamma$ (where T_Γ is the

last group of axioms of T in 3.1 restricted to the formulas in Γ), then the initial segment of M determined by $(e_i \mid i \in \mathbb{N})$ models S. Set $M \models Th(\mathbb{N})$, $M \neq \mathbb{N}$, $a \in M \setminus \mathbb{N}$, n be a natural number for S according to 2.3,(i), $b \in M$ the smallest element in M with $[a,b] \underset{*}{\to} (n+1)^n_n$. By 2.3,(i) there are $e_i, i \in \mathbb{N}$, $a \le e_i < b$, so that $(M,(e_i)_{i \in \mathbb{N}}) \models PA \cup \{c_i^2 < c_{i+1} \mid i \in \mathbb{N}\} \cup T_\Gamma$, hence $I = \{d \in M \mid \exists i \in \mathbb{N}\ d < e_i\}$ satisfies $S \cup Th_{\Pi_1}(\mathbb{N})$, but not α_n, since the existence of $b' \in I$ with $I \models [a,b'] \underset{*}{\to} (n+1)^n_n$ would contradict the minimality of b.

Remark: As Paris points out in [2], 3.7 means that in models of prefix-restricted induction in general there are not initial segments which model PA above any element; whereas 3.8 will imply that in models $\neq \mathbb{N}$ of Σ_1-induction initial segments modelling PA, other than $\mathbb{N}$, are present at all.

For the third application 3.9, on nonexistence of recursive nonstandard models, we extract from paragraph 2

3.8 Lemma: For any nonstandard $M \models PRA$ (primitive recursive arithmetic) there is a sequence $(e_i \mid i \in \mathbb{N})$, $e_i \in M$, so that $(M,(e_i)_{i \in \mathbb{N}}) \models T$ (the T of 3.1), $(e_i \mid i \in \mathbb{N})$ contained in a sequence encoded by an element of M. Thus any nonstandard model of PRA has an initial segment which is a model of PA defined by a set of Σ_o-indiscernibles.

Proof: Let S be a finite subset of T. By 3.2 and $\mathbb{N} \models \alpha$ there are $e_o,\dots,e_k \in \mathbb{N}$ so that $(\mathbb{N}, e_o,\dots,e_k) \models S$. Hence, for $M \models PRA$, $(M,e_o,\dots,e_k) \models S$, as $\mathbb{N} \subset_e M$. Using $PRA \vdash tr_o(\varphi,x) \to \varphi(x)$ for Σ_o-formulas φ the same argument as for 2.3,(ii),(1),(2) yields $(e_i \mid i \in \mathbb{N})$ as desired.

3.9 Theorem: There are no recursive nonstandard models of PRA, not even such with recursive addition.

Proof: Let $M \models PRA$, $M \neq \mathbb{N}$. By 3.8, take $e \in M$ so that $(M,((e)_i)_{i \in \mathbb{N}}) \models T$. For the initial segment I given by $(e)_i$, $I \in \mathbb{N}$, the proof of 2.3,(ii),(3) shows that for any open φ in the language of PA

(*) $I \models \exists x_1 \forall x_2 \ldots \exists x_n \varphi(x_1,\ldots,x_n)$ iff $M \models \exists x_1 < (e)_1 \forall x_2 < (e)_2 \ldots \varphi$

Set A = set of Gödel numbers of $\psi(y) \in \Sigma_o$ with $M \models \psi(e)$.

Using tr_o one realizes that there is a primitive recursive formula $\rho(z)$ containing but the parameter e so that $A = \rho^{(M,e)} \cap \mathbb{N}$. Hence by an induction on a primitive recursive formula available in PRA we get $a \in M$ with $A = \{n \in \mathbb{N} \mid M \models \underline{n} \in a\}$. Here $\underline{n} \in x$ can be taken as the formula $\exists y(\underbrace{y+\ldots+y}_{p(n)\text{-times}} = x)$, $p(n)$ the n-th prime number, a device proposed in [3]. Likewise there is a $b \in M$ so that $\mathbb{N} \setminus A = \{n \in \mathbb{N} \mid M \models n \in b\}$.

If addition were recursive on M, we could decide Th(I) as follows: For any sentence in the language of PA compute the associated Σ_o-sentence ψ on the right side of (*), then compute the formula $\underline{\psi} \in x$ and with the help of its structure and of the recursiveness of $+$ in M decide whether $M \models \psi \in a$ or $M \models \psi \in b$. But by Gödel's First Incompleteness Theorem Th(I), being a complete extension of PA, is undecidable.

Remark: The proof for the nonexistence of recursive nonstandard models M of PA given in [3] is even simpler: it just starts from a nonrecursive $A \subset \mathbb{N}$ which is arithmetically definable and thus representable as $\{n \in \mathbb{N} \mid M \models n \in a\}$ for some $a \in M$. But this representability uses an induction which is not available in PRA. Thus the reduction step from Th(I) to $Th_{\Sigma_o}((M,e))$ is decisive, yielding a nonrecursive A which is representable even in $M \models PRA$. Note that 3.9 does not hold, however, for models of the basic axioms for $0,1,+,\cdot,<,-$ plus open induction, as Sheperdson's model shows, cf. [7].

4. Generalizations of α Applying to Higher Prefixes

α is a Π_2-statement independent of PA $\cup$ $Th_{\Pi_1}(\mathbb{N})$, and we look for Π_{n+2}-formulas independent of PA $\cup$ $Th_{\Pi_{n+1}}(\mathbb{N})$ similar to α.

So we introduce generalizations α^n of α applying to Σ_n-substructures modelling PA, instead of just Σ_o-substructures. Using Σ_n-indiscernibles instead of Σ_o-indiscernibles will not do, for the proofs in paragraph 2 yield Σ_n-indiscernibles as well, if tr_o is replaced by tr_n. (This shows, by the way, that in any nonstandard $M \models$ PA we have an infinite sequence of Σ_n-indiscernibles, for any fixed $n \in \mathbb{N}$, whereas by a result of J.F. Knight ([6]) we do not have in general an infinite sequence of Σ_ω-indiscernibles). Thus we should rather try to supply sequences of Σ_o-indiscernibles so that the corresponding initial substructures be closed under Σ_n-Skolem-functions.

4.1 Definition: Let $n > 0$, tr_n the Σ_n-truth-predicate for Σ_n-formulas with two fixed variables, $g_n\colon \mathbb{N} \to \mathbb{N}$ with $g_n(a) = \mu b(\forall c \le a\ \forall d \le a\ (\exists e\, tr_n(c,d,e) \Rightarrow \exists e \le b\ tr_n(c,d,e)))$. The relation given by $g_n(a) \le b$ then becomes Π_n-definable. Let $[x,y] \overset{(n)}{\underset{*}{\to}} (z+1)^z_z$ be the formula saying: $[x,y] \to (z+1)^z_z$ plus "the required homogeneous sets X can be chosen such that $g_n(\min(X)) \le \mathrm{Card}(X)$" (instead of the special case $g_n = \mathrm{id}$ in the formula $[x,y] \underset{*}{\to} (z+1)^z_z$). Set

$$\alpha^n = \forall x \forall z \exists y [x,y] \overset{(n)}{\underset{*}{\to}} (z+1)^z_z ,$$

α^n is then a Π_{n+2}-sentence, $[x,y] \overset{(n)}{\underset{*}{\to}} (z+1)^z_z$ is a Π_n-formula.

The second proof for the independence of α given in paragraph 3 generalizes to α^n. First, there is nothing new in

<u>4.2 Lemma</u>: For each $n \in \mathbb{N}$: <u>If</u> $\varphi(x,y,z)$ is a Σ_{n+1}-formula in the language of PA so that for all $M \models PA$: (if $M \models \varphi(a,b,c)$, $c > \mathbb{N}$, then there is an initial substructure $I \subseteq_e M$ with $I \models PA$, $I \subseteq_{\Sigma_n} M$, $a \in I < b$), <u>then</u> $PA \cup Th_{\Pi_{n+1}}(\mathbb{N}) \not\vdash \forall x \forall z \exists y\, \varphi(x,y,z)$.

<u>Remark</u>: Such a φ has the main property of a definition of an indicator, this time for Σ_n-substructures which are models of PA.

Now we show that $[x,y] \overset{(n)}{\underset{*}{\rightarrow}} (z+1)^z_z$ indeed defines such a Σ_n-indicator:

<u>4.3 Lemma</u>: If $M \models PA$, $a,b \in M$, $c \in M \setminus \mathbb{N}$, $n > 0$, $M \models [a,b] \overset{(n)}{\underset{*}{\rightarrow}} (c+1)^c_c$ then there is an $I \models PA$, $I \subseteq_e M$, $I \subseteq_{\Sigma_n} M$ with $a \in I < b$.

<u>Proof</u>: $[a,b] \overset{(n)}{\underset{*}{\rightarrow}} (c+1)^c_c$ implies $[a,b] \underset{*}{\rightarrow} (c+1)^c_c$. So we have a sequence of Σ_o-indiscernibles $(e_i | i \in \mathbb{N})$, $a \le e_i < b$, as described in 2.3,(ii). Since by the proof of 2.3,(ii) this sequence arises from a homogeneous set for some partition on some $[[a,b]]^d$, the condition $[a,b] \overset{(n)}{\underset{*}{\rightarrow}} (c+1)^c_c$ yields $g^M_n(e_i) < e_{i+1}$, if we wish. In particular, as g^M_n dominates any primitive recursive function, for all $i \in \mathbb{N}$ $\langle e_o,\ldots,e_i\rangle$ (sequence-number) is smaller than e_{i+1}, all e_i being infinite by Σ_o-indiscernibility. Let I be the initial segment in M determined by $(e_i | i \in \mathbb{N})$. By 2.3,(ii) only $I \subseteq_{\Sigma_n} M$ remains to be proved. We show $I \subseteq_{\Sigma_m} M$, $m \le n$, by induction on m. By $I \subseteq_e M$ we have already $I \subseteq_{\Sigma_o} M$. Let $m < n$, $I \subseteq_{\Sigma_m} M$, $\varphi \in \Sigma_{m+1}$, $\varphi = \exists x \psi(y_1,\ldots,y_t,x)$, $\psi \in \Pi_m$. Clearly $I \models \varphi(\vec{a}) \rightarrow M \models \varphi(\vec{a})$, by $I \subseteq_{\Sigma_m} M$. So let $M \models \exists x\, \psi(a_1,\ldots,a_t,x)$, $a_j \in I$, $a_j < e_{i_j}$, $e_{i_1} < \ldots < e_{i_t}$. Equivalently $M \models \exists x \psi'(\langle a_1,\ldots,a_t\rangle,x)$, ψ' being obtained from ψ in replacing y_j by $(y)_j$. We have

$\langle a_1,\ldots,a_t\rangle < e_{i_t+1}$, $g^M_n(e_{i_t+1}) < e_{i_t+2}$, $\psi' < e_{i_t+1}$, $\psi' \in \Pi_m \subseteq \Sigma_n$, thus $M \models \exists x\, tr_n(\ulcorner\psi\urcorner, \langle a_1,\dots,a_t\rangle, x)$ entails $M \models \exists x < e_{i_t+2} tr_n(\dots)$, hence $M \models \psi'(\langle a_1,\dots,a_t\rangle, b)$ for some $b \in I$, for the same b $M \models \psi(a_1,\dots,a_t,b)$, and by induction hypothesis this holds in I.

Finally, to render 4.2 and 4.3 useful, we need

4.4 Lemma: For all $n > 0$ $\mathbb{N} \models \alpha^n$.
As to its proof, we only remark that in the proof of $\mathbb{N} \models \alpha$ the homogeneous sets X can be taken from an infinite set, so it is not difficult to achieve $f(\min(X)) \leq \mathrm{Card}(X)$ for any fixed function $f:\mathbb{N}\to\mathbb{N}$.

Combining 4.2 - 4.4 we obtain

4.5 Theorem: For all $n > 0$ α^n is a true Π_{n+2}-sentence independent of $PA \cup Th_{\Pi_{n+1}}(\mathbb{N})$. Moreover, if $M \models PA$, $M \neq \mathbb{N}$, there is an $I \subseteq_e M$, $I \models PA$, so that $I \subseteq_{\Sigma_n} M$ and $I \models \neg\alpha^n$.

Remark: For the first part of 4.5, 4.3 is only needed in the special case $M \models Th(\mathbb{N})$.

4.6 Corollary: For all $n \in \mathbb{N}$ any nonstandard model of $Th(\mathbb{N})$ has an infinity of initial Σ_n-substructures satisfying pairwise different complete extensions of PA

Remark: Ehrenfeucht and D. Jensen proved in [3] for $n = 0$ the stronger result that a nonstandard model of PA has $2^{\aleph_0}$ distinct completions of PA satisfied by its initial substructures. Note that for $n = 0$ 4.6 immediately generalizes to models of $PA \cup Th_{\Pi_1}(\mathbb{N})$, with regard to the result by Gaifman quoted in the remark following 3.4.

References

[1] J. Paris, L. Harrington: A Mathematical Incompleteness in Peano Arithmetic, in: Handbook of Mathematical Logic, ed. Jon Barwise, 1133-1142.

[2] J. Paris: Some Independence Results for Peano Arithmetic, The Journal of Symbolic Logic, vol. 43, No. 4, 1978, 725-731.

[3] A. Ehrenfeucht, D. Jensen: Some problems in elementary arithmetics, Fundamenta Mathematical, XCII, 1976, 223-245.

[4] H. Friedman: Countable Models of Set Theories, in: Proc. of the 1971 Cambridge Summer School in Mathematical Logic, Springer Lecture Notes, in Math., vol. 337, 539-573.

[5] H. Gaifman: A Note on Models and Submodels of Arithmetic, Proc. of the Conference in Mathematical Logic, London 1970, Springer Lecture Notes in Mathematics, vol. 255, 128-144.

[6] J. F. Knight: Types Omitted in Uncountable Models of Arithmetic, The Journal of Symbolic Logic, vol. 40, 1975, 317-320.

[7] J.C. Sheperdson: A Nonstandard Model for a Free Variable Fragment of Number Theory, Bull. de l'acadêmie Polonaise des Sciences, Sêr. Math., Astr., Phys., vol. XII, No. 2, 1964, 79-86.

Vol. 700: Module Theory, Proceedings, 1977. Edited by C. Faith and S. Wiegand. X, 239 pages. 1979.

Vol. 701: Functional Analysis Methods in Numerical Analysis, Proceedings, 1977. Edited by M. Zuhair Nashed. VII, 333 pages. 1979.

Vol. 702: Yuri N. Bibikov, Local Theory of Nonlinear Analytic Ordinary Differential Equations. IX, 147 pages. 1979.

Vol. 703: Equadiff IV, Proceedings, 1977. Edited by J. Fábera. XIX, 441 pages. 1979.

Vol. 704: Computing Methods in Applied Sciences and Engineering, 1977, I. Proceedings, 1977. Edited by R. Glowinski and J. L. Lions. VI, 391 pages. 1979.

Vol. 705: O. Forster und K. Knorr, Konstruktion verseller Familien kompakter komplexer Räume. VII, 141 Seiten. 1979.

Vol. 706: Probability Measures on Groups, Proceedings, 1978. Edited by H. Heyer. XIII, 348 pages. 1979.

Vol. 707: R. Zielke, Discontinuous Čebyšev Systems. VI, 111 pages. 1979.

Vol. 708: J. P. Jouanolou, Equations de Pfaff algébriques. V, 255 pages. 1979.

Vol. 709: Probability in Banach Spaces II. Proceedings, 1978. Edited by A. Beck. V, 205 pages. 1979.

Vol. 710: Séminaire Bourbaki vol. 1977/78, Exposés 507–524. IV, 328 pages. 1979.

Vol. 711: Asymptotic Analysis. Edited by F. Verhulst. V, 240 pages. 1979.

Vol. 712: Equations Différentielles et Systèmes de Pfaff dans le Champ Complexe. Edité par R. Gérard et J.-P. Ramis. V, 364 pages. 1979.

Vol. 713: Séminaire de Théorie du Potentiel, Paris No. 4. Edité par F. Hirsch et G. Mokobodzki. VII, 281 pages. 1979.

Vol. 714: J. Jacod, Calcul Stochastique et Problèmes de Martingales. X, 539 pages. 1979.

Vol. 715: Inder Bir S. Passi, Group Rings and Their Augmentation Ideals. VI, 137 pages. 1979.

Vol. 716: M. A. Scheunert, The Theory of Lie Superalgebras. X, 271 pages. 1979.

Vol. 717: Grosser, Bidualräume und Vervollständigungen von Banachmoduln. III, 209 pages. 1979.

Vol. 718: J. Ferrante and C. W. Rackoff, The Computational Complexity of Logical Theories. X, 243 pages. 1979.

Vol. 719: Categorial Topology, Proceedings, 1978. Edited by H. Herrlich and G. Preuß. XII, 420 pages. 1979.

Vol. 720: E. Dubinsky, The Structure of Nuclear Fréchet Spaces. V, 187 pages. 1979.

Vol. 721: Séminaire de Probabilités XIII. Proceedings, Strasbourg, 1977/78. Edité par C. Dellacherie, P. A. Meyer et M. Weil. VII, 647 pages. 1979.

Vol. 722: Topology of Low-Dimensional Manifolds. Proceedings, 1977. Edited by R. Fenn. VI, 154 pages. 1979.

Vol. 723: W. Brandal, Commutative Rings whose Finitely Generated Modules Decompose. II, 116 pages. 1979.

Vol. 724: D. Griffeath, Additive and Cancellative Interacting Particle Systems. V, 108 pages. 1979.

Vol. 725: Algèbres d'Opérateurs. Proceedings, 1978. Edité par P. de la Harpe. VII, 309 pages. 1979.

Vol. 726: Y.-C. Wong, Schwartz Spaces, Nuclear Spaces and Tensor Products. VI, 418 pages. 1979.

Vol. 727: Y. Saito, Spectral Representations for Schrödinger Operators With Long-Range Potentials. V, 149 pages. 1979.

Vol. 728: Non-Commutative Harmonic Analysis. Proceedings, 1978. Edited by J. Carmona and M. Vergne. V, 244 pages. 1979.

Vol. 729: Ergodic Theory. Proceedings, 1978. Edited by M. Denker and K. Jacobs. XII, 209 pages. 1979.

Vol. 730: Functional Differential Equations and Approximation of Fixed Points. Proceedings, 1978. Edited by H.-O. Peitgen and H.-O. Walther. XV, 503 pages. 1979.

Vol. 731: Y. Nakagami and M. Takesaki, Duality for Crossed Products of von Neumann Algebras. IX, 139 pages. 1979.

Vol. 732: Algebraic Geometry. Proceedings, 1978. Edited by K. Lønsted. IV, 658 pages. 1979.

Vol. 733: F. Bloom, Modern Differential Geometric Techniques in the Theory of Continuous Distributions of Dislocations. XII, 206 pages. 1979.

Vol. 734: Ring Theory, Waterloo, 1978. Proceedings, 1978. Edited by D. Handelman and J. Lawrence. XI, 352 pages. 1979.

Vol. 735: B. Aupetit, Propriétés Spectrales des Algèbres de Banach. XII, 192 pages. 1979.

Vol. 736: E. Behrends, M-Structure and the Banach-Stone Theorem. X, 217 pages. 1979.

Vol. 737: Volterra Equations. Proceedings 1978. Edited by S.-O. Londen and O. J. Staffans. VIII, 314 pages. 1979.

Vol. 738: P. E. Conner, Differentiable Periodic Maps. 2nd edition, IV, 181 pages. 1979.

Vol. 739: Analyse Harmonique sur les Groupes de Lie II. Proceedings, 1976–78. Edited by P. Eymard et al. VI, 646 pages. 1979.

Vol. 740: Séminaire d'Algèbre Paul Dubreil. Proceedings, 1977–78. Edited by M.-P. Malliavin. V, 456 pages. 1979.

Vol. 741: Algebraic Topology, Waterloo 1978. Proceedings. Edited by P. Hoffman and V. Snaith. XI, 655 pages. 1979.

Vol. 742: K. Clancey, Seminormal Operators. VII, 125 pages. 1979.

Vol. 743: Romanian-Finnish Seminar on Complex Analysis. Proceedings, 1976. Edited by C. Andreian Cazacu et al. XVI, 713 pages. 1979.

Vol. 744: I. Reiner and K. W. Roggenkamp, Integral Representations. VIII, 275 pages. 1979.

Vol. 745: D. K. Haley, Equational Compactness in Rings. III, 167 pages. 1979.

Vol. 746: P. Hoffman, τ-Rings and Wreath Product Representations. V, 148 pages. 1979.

Vol. 747: Complex Analysis, Joensuu 1978. Proceedings, 1978. Edited by I. Laine, O. Lehto and T. Sorvali. XV, 450 pages. 1979.

Vol. 748: Combinatorial Mathematics VI. Proceedings, 1978. Edited by A. F. Horadam and W. D. Wallis. IX, 206 pages. 1979.

Vol. 749: V. Girault and P.-A. Raviart, Finite Element Approximation of the Navier-Stokes Equations. VII, 200 pages. 1979.

Vol. 750: J. C. Jantzen, Moduln mit einem höchsten Gewicht. III, 195 Seiten. 1979.

Vol. 751: Number Theory, Carbondale 1979. Proceedings. Edited by M. B. Nathanson. V, 342 pages. 1979.

Vol. 752: M. Barr, *-Autonomous Categories. VI, 140 pages. 1979.

Vol. 753: Applications of Sheaves. Proceedings, 1977. Edited by M. Fourman, C. Mulvey and D. Scott. XIV, 779 pages. 1979.

Vol. 754: O. A. Laudal, Formal Moduli of Algebraic Structures. III, 161 pages. 1979.

Vol. 755: Global Analysis. Proceedings, 1978. Edited by M. Grmela and J. E. Marsden. VII, 377 pages. 1979.

Vol. 756: H. O. Cordes, Elliptic Pseudo-Differential Operators – An Abstract Theory. IX, 331 pages. 1979.

Vol. 757: Smoothing Techniques for Curve Estimation. Proceedings, 1979. Edited by Th. Gasser and M. Rosenblatt. V, 245 pages. 1979.

Vol. 758: C. Năstăsescu and F. Van Oystaeyen; Graded and Filtered Rings and Modules. X, 148 pages. 1979.

Vol. 759: R. L. Epstein, Degrees of Unsolvability: Structure and Theory. XIV, 216 pages. 1979.

Vol. 760: H.-O. Georgii, Canonical Gibbs Measures. VIII, 190 pages. 1979.

Vol. 761: K. Johannson, Homotopy Equivalences of 3-Manifolds with Boundaries. 2, 303 pages. 1979.

Vol. 762: D. H. Sattinger, Group Theoretic Methods in Bifurcation Theory. V, 241 pages. 1979.

Vol. 763: Algebraic Topology, Aarhus 1978. Proceedings, 1978. Edited by J. L. Dupont and H. Madsen. VI, 695 pages. 1979.

Vol. 764: B. Srinivasan, Representations of Finite Chevalley Groups. XI, 177 pages. 1979.

Vol. 765: Padé Approximation and its Applications. Proceedings, 1979. Edited by L. Wuytack. VI, 392 pages. 1979.

Vol. 766: T. tom Dieck, Transformation Groups and Representation Theory. VIII, 309 pages. 1979.

Vol. 767: M. Namba, Families of Meromorphic Functions on Compact Riemann Surfaces. XII, 284 pages. 1979.

Vol. 768: R. S. Doran and J. Wichmann, Approximate Identities and Factorization in Banach Modules. X, 305 pages. 1979.

Vol. 769: J. Flum, M. Ziegler, Topological Model Theory. X, 151 pages. 1980.

Vol. 770: Séminaire Bourbaki vol. 1978/79 Exposés 525–542. IV, 341 pages. 1980.

Vol. 771: Approximation Methods for Navier-Stokes Problems. Proceedings, 1979. Edited by R. Rautmann. XVI, 581 pages. 1980.

Vol. 772: J. P. Levine, Algebraic Structure of Knot Modules. XI, 104 pages. 1980.

Vol. 773: Numerical Analysis. Proceedings, 1979. Edited by G. A. Watson. X, 184 pages. 1980.

Vol. 774: R. Azencott, Y. Guivarc'h, R. F. Gundy, Ecole d'Eté de Probabilités de Saint-Flour VIII-1978. Edited by P. L. Hennequin. XIII, 334 pages. 1980.

Vol. 775: Geometric Methods in Mathematical Physics. Proceedings, 1979. Edited by G. Kaiser and J. E. Marsden. VII, 257 pages. 1980.

Vol. 776: B. Gross, Arithmetic on Elliptic Curves with Complex Multiplication. V, 95 pages. 1980.

Vol. 777: Séminaire sur les Singularités des Surfaces. Proceedings, 1976-1977. Edited by M. Demazure, H. Pinkham and B. Teissier. IX, 339 pages. 1980.

Vol. 778: SK1 von Schiefkörpern. Proceedings, 1976. Edited by P. Draxl and M. Kneser. II, 124 pages. 1980.

Vol. 779: Euclidean Harmonic Analysis. Proceedings, 1979. Edited by J. J. Benedetto. III, 177 pages. 1980.

Vol. 780: L. Schwartz, Semi-Martingales sur des Variétés, et Martingales Conformes sur des Variétés Analytiques Complexes. XV, 132 pages. 1980.

Vol. 781: Harmonic Analysis Iraklion 1978. Proceedings 1978. Edited by N. Petridis, S. K. Pichorides and N. Varopoulos. V, 213 pages. 1980.

Vol. 782: Bifurcation and Nonlinear Eigenvalue Problems. Proceedings, 1978. Edited by C. Bardos, J. M. Lasry and M. Schatzman. VIII, 296 pages. 1980.

Vol. 783: A. Dinghas, Wertverteilung meromorpher Funktionen in ein- und mehrfach zusammenhängenden Gebieten. Edited by R. Nevanlinna and C. Andreian Cazacu. XIII, 145 pages. 1980.

Vol. 784: Séminaire de Probabilités XIV. Proceedings, 1978/79. Edited by J. Azéma and M. Yor. VIII, 546 pages. 1980.

Vol. 785: W. M. Schmidt, Diophantine Approximation. X, 299 pages. 1980.

Vol. 786: I. J. Maddox, Infinite Matrices of Operators. V, 122 pages. 1980.

Vol. 787: Potential Theory, Copenhagen 1979. Proceedings, 1979. Edited by C. Berg, G. Forst and B. Fuglede. VIII, 319 pages. 1980.

Vol. 788: Topology Symposium, Siegen 1979. Proceedings, 1979. Edited by U. Koschorke and W. D. Neumann. VIII, 495 pages. 1980.

Vol. 789: J. E. Humphreys, Arithmetic Groups. VII, 158 pages. 1980.

Vol. 790: W. Dicks, Groups, Trees and Projective Modules. IX, 127 pages. 1980.

Vol. 791: K. W. Bauer and S. Ruscheweyh, Differential Operators for Partial Differential Equations and Function Theoretic Applications. V, 258 pages. 1980.

Vol. 792: Geometry and Differential Geometry. Proceedings, 1979. Edited by R. Artzy and I. Vaisman. VI, 443 pages. 1980.

Vol. 793: J. Renault, A Groupoid Approach to C*-Algebras. III, 160 pages. 1980.

Vol. 794: Measure Theory, Oberwolfach 1979. Proceedings 1979. Edited by D. Kölzow. XV, 573 pages. 1980.

Vol. 795: Séminaire d'Algèbre Paul Dubreil et Marie-Paule Malliavin. Proceedings 1979. Edited by M. P. Malliavin. V, 433 pages. 1980.

Vol. 796: C. Constantinescu, Duality in Measure Theory. IV, 197 pages. 1980.

Vol. 797: S. Mäki, The Determination of Units in Real Cyclic Sextic Fields. III, 198 pages. 1980.

Vol. 798: Analytic Functions, Kozubnik 1979. Proceedings. Edited by J. Ławrynowicz. X, 476 pages. 1980.

Vol. 799: Functional Differential Equations and Bifurcation. Proceedings 1979. Edited by A. F. Izé. XXII, 409 pages. 1980.

Vol. 800: M.-F. Vignéras, Arithmétique des Algèbres de Quaternions. VII, 169 pages. 1980.

Vol. 801: K. Floret, Weakly Compact Sets. VII, 123 pages. 1980.

Vol. 802: J. Bair, R. Fourneau, Etude Géometrique des Espaces Vectoriels II. VII, 283 pages. 1980.

Vol. 803: F.-Y. Maeda, Dirichlet Integrals on Harmonic Spaces. X, 180 pages. 1980.

Vol. 804: M. Matsuda, First Order Algebraic Differential Equations VII, 111 pages. 1980.

Vol. 805: O. Kowalski, Generalized Symmetric Spaces. XII, 187 pages. 1980.

Vol. 806: Burnside Groups. Proceedings, 1977. Edited by J. L. Mennicke. V, 274 pages. 1980.

Vol. 807: Fonctions de Plusieurs Variables Complexes IV. Proceedings, 1979. Edited by F. Norguet. IX, 198 pages. 1980.

Vol. 808: G. Maury et J. Raynaud, Ordres Maximaux au Sens de K. Asano. VIII, 192 pages. 1980.

Vol. 809: I. Gumowski and Ch. Mira, Recurences and Discrete Dynamic Systems. VI, 272 pages. 1980.

Vol. 810: Geometrical Approaches to Differential Equations. Proceedings 1979. Edited by R. Martini. VII, 339 pages. 1980.

Vol. 811: D. Normann, Recursion on the Countable Functionals. VIII, 191 pages. 1980.

Vol. 812: Y. Namikawa, Toroidal Compactification of Siegel Spaces VIII, 162 pages. 1980.

Vol. 813: A. Campillo, Algebroid Curves in Positive Characteristic V, 168 pages. 1980.

Vol. 814: Séminaire de Théorie du Potentiel, Paris, No. 5. Proceedings. Edited by F. Hirsch et G. Mokobodzki. IV, 239 pages. 1980.

Vol. 815: P. J. Slodowy, Simple Singularities and Simple Algebraic Groups. XI, 175 pages. 1980.

Vol. 816: L. Stoica, Local Operators and Markov Processes. VIII, 104 pages. 1980.